LUFTWAFFE TRAN~~SPORT UNITS~~

"I have no wish to belittle in any way the sacrifices made by the air transport forces or the feats accomplished by them [but] it might have been better if the Demyansk action had failed completely in its initial phase. This would have forced German leaders to order the 2.Armeekorps to attempt a breakthrough towards the west and to initiate simultaneous small-scale counter-attacks in order to relieve the pressure at Demyansk. At the same time, German leaders would have been forced to realise that the personnel and material resources available were inadequate to assure fulfilment by air of the entire supply requirements of an Army unit numbering 100,000 men. Whether this would have had any effect on the decisions made later in connection with Stalingrad and North Africa is, of course, another question."

From a post-war report by Generalmajor A.D. Freidrich-Wilhelm Morzik,
formerly Wehrmacht Lufttransportchef OKL

"I want Transporters, Transporters and more Transporters!"

Adolf Hitler to Generalfeldmarschall Erhard Milch, 4 February 1943

Pre-War Development

The history of the *Luftwaffe's* entire air transport force during the Second World War is inseparably linked to the Junkers Ju 52/3m. This aircraft, although obsolescent long before the war began, remained in production as an all-purpose transport and training machine until October 1944 and fully justified that measure by showing such astonishing reliability and longevity that it was still the *Luftwaffe's* standard transport aircraft in 1945.

During the period between 1933 and 1935, when most German-designed aircraft were presented to the world as civil or sports types, the Ju 52/3m proved to be an extremely reliable passenger aircraft and was employed in some numbers by the German airline *Deutsche Lufthansa*. With the creation of the *Luftwaffe* in 1935, the aircraft emerged as a bomber and subsequently took part in several

BELOW: This Lufthansa machine, W.Nr. 4020, was named after the First World War fighter pilot Gustav Dörr, who flew as a Leutnant with Jasta 45 and was credited with 35 victories.

ABOVE: An example of a Ju 52/3m ge in Lufthansa markings. Below the swastika banner on the tail is the aircraft's full designation and the W.Nr. 4041. This particular machine was named after the First World War fighter pilot Werner Voss, credited with 48 victories, and carried his name under the Lufthansa legend on the nose. The standard Lufthansa scheme was an overall light grey or silver painted finish with black areas on the nose, engine areas and wheel spats. Although shown here with the civil registration D-3131, the aircraft was later re-registered as D-ARAM.

bombing missions with the *Legion Condor* during the Spanish Civil War. From 1937 however, with the development of more modern bomber aircraft, the Ju 52/3m was gradually withdrawn from this role and replaced by the new Do 17 and He 111 types.

Although some Ju 52/3ms were then handed over to various special staff and courier units, the majority found their way into advanced pilot and instrument schools where they were required to increase training capacity commensurate with the growth of the *Luftwaffe*. Although already obsolescent, production of the Ju 52/3m was maintained in order to provide the *Luftwaffe* with an aircraft which could be employed as a transport for the growing German paratroop, or *Fallschirmjäger*, and *Luftlande*, or air-landing forces[1]. That the rapid commitment of troops by air was both desirable and possible had already been amply demonstrated at the beginning of the Spanish Civil War when large Moroccan troop units were flown into Spain from Tetuan, in Morocco, by Ju 52/3ms. This was an undertaking which, as well as having a decisive effect on the ensuing land battle, also confirmed the suitability of the Ju 52/3m for such purposes.

The Ju 52/3m was finally abandoned as a bomber in 1938, by which time most bomber *Gruppen* had converted to more modern types. Meanwhile, in October 1937, IV./KG 152 'Hindenburg' at Fürstenwalde/Spree was redesignated KGrzbV I, the abbreviation zbV standing for *zur besonderen Verwendung*, i.e. for special employment; the special employment envisaged being the dropping of parachutists and the transportation of airborne troops. The unit had a strength of 39 Ju 52/3ms and later took up headquarters at Burg and Gardelegen. On 1 August, a second *Gruppe*, KGrzbV II, was formed at Brandenburg-Briest and Stendal, both units being controlled by the parachute and airborne troops command, 7. *Fliegerdivision*.

The air transport units played little part in the *Anschluss*, the annexation of Austria, which

BELOW: During the Spanish Civil War, when General Franco requested aircraft to transport army units in Morocco to Spain, Hitler immediately promised 20 Ju 52/3ms for the task. Here, Moroccan soldiers are shown waiting to board their aircraft for the flight to Spain. Some 10,000 Moroccan troops were transferred to Spain, but as the soldiers suffered badly from air sickness, the Ju 52/3ms employed for this task had to be cleaned quite frequently.

---

1. The Fallschirmjäger were intended to parachute into action whereas the troops of the Luftlande Division were airlifted into battle. The Luftlande troops belonged to the Army, and although the Army also had paratroops, the majority comprised Luftwaffe personnel.

began on 12 March 1938, although several formations were called up from the schools and, on the morning of 13 March, they flew over Austria in *Gruppe* formation dropping leaflets over the larger towns and cities. At the same time, a parachute battalion was transported to Graz in an air-landing operation. A week later, a temporary *Transport Gruppe* from the instrument flight school at Wesendorf was transferred to Garz, an airfield on Usedom Island in the Baltic, with orders to transport the entire ground organisation of a *Jagdgeschwader* to East Prussia in preparation for the occupation of Memel in Lithuania.

By the summer of 1938, the parachute forces had been built up into a powerful and immediately available force, and since armed resistance to the planned German occupation of the Sudetenland could not be ruled out, 7. *Fliegerdivision* made plans for a combined paratroop and air-landing operation behind the Czech bunker line near Freudenthal. The transport units therefore augmented their standard training with special instruction in the landing of fully laden Ju 52/3ms on auxiliary airfields or in open terrain under purposely difficult conditions. Each pilot was taught first to land by himself, then as a member of a flight, and finally as a member of a *Staffel*, in areas where space limitations required the approaching formation to manoeuvre into single file, land, circle back down the field next to the landing strip, unload, taxi round to the runway, and take off in the shortest possible time to make way for the next unit. These practice flights were supplemented by map exercises and illustrated lectures to emphasise the point that the success of the whole undertaking depended on the smooth functioning of the air-landings.

LEFT: Ju 52/3ms in the standard pre-war uppersurface scheme of green RLM 70/71 but with black, instead of blue 65, undersurfaces. The small Balkenkreuz visible on the fuselage of the aircraft in the centre of the picture and the swastika banner in red, black and white over the tail is typical of the period between March 1938 and March 1939 when Germany occupied Austria, annexed the Sudetenland, and finally occupied the whole of Czechoslovakia. The aircraft shown here were operating under the control of KGzbV 2.

In the event, however, the occupation of the Sudetenland was accomplished peaceably, but pilots who later took part in the airlifts and operations to supply encircled troops on the Russian Front found this thorough training extremely valuable. Had this training been extended so that the transport units practised to meet the needs of the Army, the exercise would have proved invaluable during the later large-scale airlift and supply operations in the East. As it was, this experience was not gained until later during the war when the use of combined parachute and air-landing operations was overshadowed by the growing need for simple air-supply missions. By that time, however, the expertise had to be paid for dearly in terms of personnel and equipment.

In the summer of 1939, the existing KGzbV were redesignated to form the first two *Gruppen* of *Kampfgeschwader* zbV 1; KGrzbV I and II being redesignated I. and II./KGzbV 1 respectively. At the same time, a thrifty policy of establishing zbV units at the *C-Schulen*, the advanced twin-engine schools, was initiated. This was carried out in such a way that for any special parachute or airborne operation, sufficient aircraft and instructor crews could be withdrawn from the schools to form two further *Gruppen* but returned to the schools during the periods between such operations. Thus, there now existed two Ju 52/3m *Gruppen*, I. and II./KGzbV 1 and, existing on paper but not actually called up for operational employment, the two further *Gruppen* at the *C-Schulen* which, on mobilisation, were to be designated III. and IV./KGzbV 1.

At the end of August 1939, three further transport *Gruppen* were established: KGrzbV 9 and I. and II./KGzbV 172. KGrzbV 9 consisted of *C-Schule* personnel and was to be employed in transporting heavy weapons, while I. and II./KGzbV 172 were raised from Berlin Air Liaison Staff and *Deutsche Lufthansa* personnel. The establishment of a transport *Geschwader* was now settled at 217 aircraft: the *Geschwader Stab* was to have five aircraft; each *Gruppen* five; and each of the four *Gruppen* was to consist of four *Staffeln* of 12 aircraft each. As each machine could carry between 15 and 18 men, a *Gruppe* could lift a parachute or airborne battalion and a transport *Geschwader* could carry a regiment.

ABOVE: The Regierungsstaffel, or Government Staffel, was a permanently established unit of courier aircraft employed for the benefit of Adolf Hitler and high-ranking party and military officials. The Ju 52/3m in the centre of this photograph, probably taken in 1936, is the Führermaschine, also known as the Führer-Ju, which was Hitler's personal transport. This aircraft, W.Nr. 4021, was registered as D-2600 and carried the name 'Immelmann'. It remained Hitler's personal transport until replaced by an Fw 200 in October 1939. The Ju 52/3m on the right, registered D-AJIM and named 'Hermann Göring', is another machine of the Regierungsstaffel and was assigned to the Reich's War Minister, Feldmarschall Werner von Blomberg, until he was dismissed in February 1937.

BELOW: The standard Ju 52/3m livery shown from above on W.Nr. 4019 when registered as D-2468. A swastika banner has yet to be applied to the vertical tail surfaces and the machine was later re-registered as D-AFIR.

LEFT: Generaloberst Hermann Göring visiting Luftwaffe units based at Lechfeld in 1936. The Ju 52/3m in the background is one of three such aircraft allocated to Göring, all of which featured large areas of red and white with striped ailerons and flaps. Later, these machines each acquired black engine cowlings and nacelles with the nose also in black and with Lufthansa's characteristic curving demarcation line.

BELOW: Hermann Göring shown in the cockpit of 'Manfred von Richthofen II', W.Nr. 4066, which carried the registration D-ABAQ. Later, when the Ju 52/3ms of the Regierungsstaffel were replaced by Fw 200s, they were returned to Lufthansa for ordinary commercial use and were repainted with different markings. W.Nr. 4066 eventually saw service with the Luftwaffe and, with the Stammkennzeichen GC+AE, was lost in January 1943.

LEFT: Two views of the luxurious interior of one of Göring's Ju 52/3ms.

1939-1943

LEFT AND BELOW: Two views of Generaloberst Göring's Ju 52/3m, W.Nr. 4022. This aircraft was allocated exclusively for Göring's use and was named 'Manfred von Richthofen'. His other two machines, W.Nr. 4066 registered D-ABAQ, and W.Nr. 4069 registered D-ABIK, were named 'Manfred von Richthofen II' and 'Manfred von Richthofen III' and were used by senior staff of the RLM as well as by Göring himself. Note that in the photograph (BELOW), the aircraft is shown with a tailskid and wheel spats removed.

A plan view of D-2527 showing the demarcation between the red and white livery on the wings and the upper fuselage. The undersurfaces were identical except that the white blaze on the fuselage was omitted.

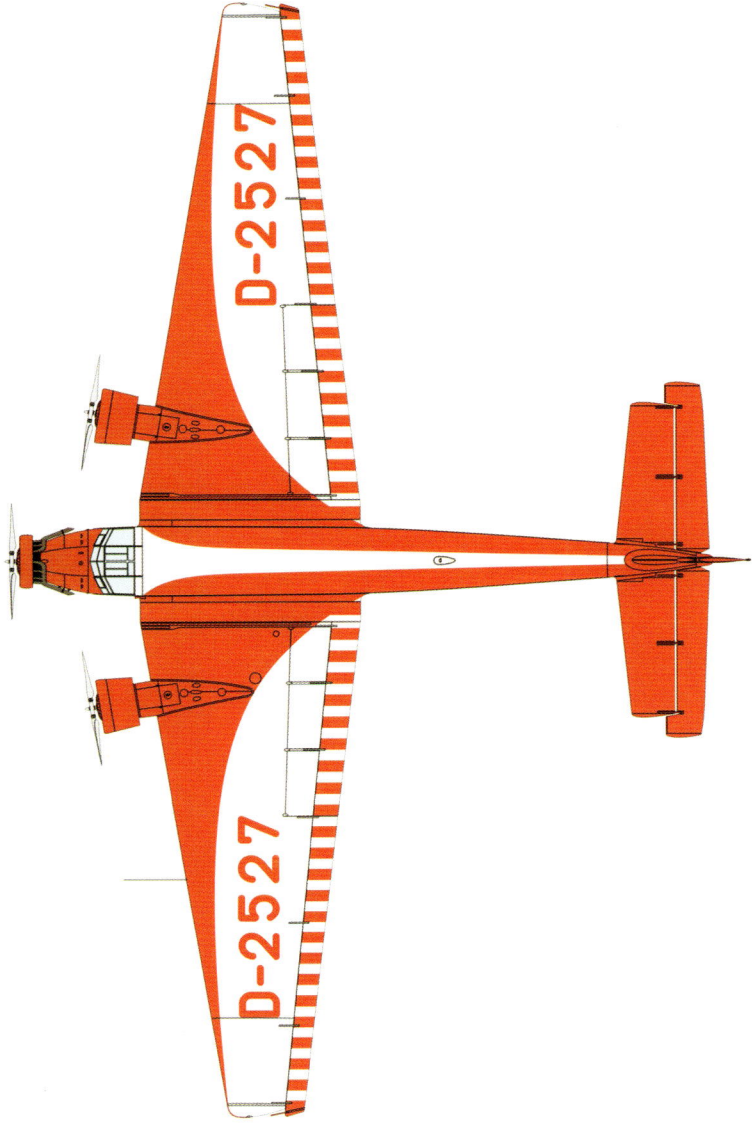

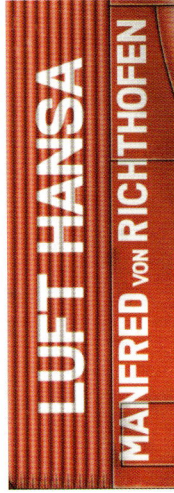

LUFT HANSA

MANFRED VON RICHTHOFEN

Detail of nose lettering

Junkers Ju 52/3m 'Manfred von Richthofen' of the Regierungsstaffel

This aircraft, W.Nr. 4022, was registered as D-2527 and featured an elaborate red and white livery in commemoration of Rittmeister Manfred von Richthofen and his red Fokker Dr.I triplanes of First World War fame. The machine is shown here with the small, circular window aft of the main door painted over. Note also the position of the swastika on the rudder and that the natural metal propeller blades were black on their rear faces.

The Polish Campaign

In September 1939, the *Luftwaffe* possessed some 400 Ju 52/3ms equipped to transport its parachute and airborne forces and on 1 September, KGzbV 1 was transferred to its operational and front-line airfields in Silesia. Here the various *Gruppen* met up with paratroop units that had marched to the assembly points several days earlier. Directly after hostilities began, negotiations were opened between Germany and Poland and the possibility of a truce, perhaps followed by orders that all troops should remain in the positions they had reached, had to be considered. If that had occurred, then parachute troops dropped along the 1914 Polish-German border might have considerably influenced Germany's position in the event of a peace conference.

During the first few days of the campaign, 7. *Fliegerdivision* issued alert orders to all the transport and parachute units under its command, but as the truce negotiations were unsuccessful, the entire plan was dropped and air transport was employed only to lift part of the 22. *Infanterie Division* from Silesia to the Lodz area of Poland. Air transport units were also required to supply armoured spearheads and operational flying units on the southern sector of the Polish front; this was the first occasion on which Ju 52/3m aircraft were used in purely re-supply missions. Losses were very light, and at the end of the campaign all units except I. and II./KGzbV 1 were returned to the *C-Schulen,* on the understanding that the *Gruppenstab* and *Staffel* HQ sections were to be kept together at the same school so that the original units could be quickly reassembled should the need arise.

During the campaign in Poland, the transport units were employed to airlift troops and to supply armoured spearheads and operational flying units on the southern sector of the front. The Ju 52/3m seen here belonged to an unknown unit and was photographed near Lublin in Poland in late 1939. Note that with the exception of the aircraft letter E on the fuselage, all operational and national markings have been oversprayed.

Operations in the West

Operation *'Weserübung'*, the almost simultaneous occupation of Denmark and Norway in April 1940, was the first major test of the capabilities of the air transport units, for whereas their role in Poland had been comparatively unimportant, the success of a whole campaign was now dependent upon the accomplishment of the missions to which they were assigned.

For the invasion of Norway, it was essential that the Danish airfields at Aalborg and Vordingborg were rapidly occupied as both were required as bases from which to launch parachute operations to capture airfields in Norway. The air aspect of the attack was entrusted to X. *Fliegerkorps*, and by 5 April, the following transport units had transferred to their jumping-off airfields:

Stab and I./KGzbV 1	based at Ütersen
II/KGzbV 1 (minus 7. and 8. *Staffel*)	based at Schleswig
7./KGzbV 1	based at Stade
8./KGzbV 1	based at Ütersen
III./KGzbV 1	based at Hagenow
IV./KGzbV 1	based at Hagenow

Two or three days before *'Weserübung'* was due to be launched, the air transport units moved from their home airfields to assembly bases, and the training units were alerted to stand by at their assembly airfields. The parachute and air-landing units arrived at their appointed take-off bases during 8 April. Late that evening, the codeword for the operation was received and the unit commanders removed their operational orders from their sealed envelopes; *Unternehmen 'Weserübung'* was to begin at 05.30 hrs the next day.

The morning of 9 April dawned cloudless but hazy and although morning fog over the sea was anticipated, it was reported as being lower than the altitude at which the transports would fly. The first wave of Ju 52/3ms therefore took off on schedule as soon as it began to grow light. The Danish and Norwegian governments had already received ultimatums advising them of the pending German military occupation and, during the flight, the German forces were informed that while Denmark was prepared to accept occupation, Norway was determined to resist. Consequently, the landings in Denmark were completed smoothly and without opposition. Accompanied by three signals aircraft of X. *Fliegerkorps* and protected by two *Staffeln* of Bf 110s from I./ZG 76, the Ju 52/3ms of 8./KGzbV 1 dropped the men of 4./FJR 1 (4th Company, *Fallschirmjäger Regiment* 1) near Aalborg-West to protect the aircraft of I./KGzbV 1 as they landed the troops of 1. *Bataillon*, XXXI *Armeekorps*. As soon as the area had been secured, I., II. and III./KGzbV 1 landed III./*Infanterie Regiment* 159 at both Aalborg airfields, which could then be used as emergency fields to refuel the transport aircraft flying to and from Norway. Equally important was the transport of ground staff and ammunition, etc, for the heavy fighters of I./ZG 76 which were to protect transport aircraft in flight over the Skagerrak.

Meanwhile, at 08.45 hrs, 12 aircraft from 7./KGzbV 1 dropped 3./FJR 1 on Stavanger-Sola airfield in Norway. These aircraft were accompanied by two signals aircraft from V. *Fliegerkorps* which were to set up a W/T and D/F station, and were escorted by a *Schwarm* of Bf 110s from I./ZG 76. The Norwegian defenders had set up barbed-wire barriers to prevent the field from being used, but these were quickly torn down by a group of paratroops who cleared a landing strip and secured the airfield for the Ju 52/3ms of KGrzbV 104 which landed a few minutes later with the troops of *Stab*, I. and II./*Infanterie Regiment* 193. Shortly afterwards, three Ju 52/3ms from KGrzbV 101 landed with drums of fuel for the heavy fighter *Schwarm* of I./ZG 76 which had protected the landing.

The most important phase of the operation, however, was the air landing at Oslo's Fornebu airport. This objective had to be seized as soon as possible as it was required as a base from which bomber and fighter aircraft would provide support and air cover for ground forces moving inland and also as a main unloading point for the supply and reinforcement of German forces at Narvik. Any delay or setback in the capture of Fornebu could have seriously jeopardised operations in central Norway and, indeed, the whole undertaking.

Again the plan called for a multi-wave attack; a paratroop drop followed by air-landed troops. The first wave of aircraft, comprising 3. and 6./KGzbV 1 under the command of *Major* Karl Drewes, was over the Skagerrak en route for Oslo-Fornebu when they ran into a fog bank which extended far above the altitude at which they had been ordered to fly. At first, the fog was patchy and fairly scattered, so even the less experienced pilots had no difficulty in remaining together. But as the fog grew thicker, visibility was reduced to only a few metres and formation flying was made impossible. The inexperienced pilots, making up a quarter of the formation, were not capable of going on much further, yet the weather showed no signs of improvement. In view of the importance of the mission, *Major* Drewes delayed the decision to turn back as long as possible, but after two pilots lost control of their aircraft and crashed with crews and paratroops into the water, Drewes signalled X. *Fliegerkorps* that he was returning. X. *Fliegerkorps* then ordered all the remaining aircraft of the first and second waves to turn back.

However, the second wave of aircraft had all been drawn from the training units and their instructor pilots were perfectly capable of coping with the adverse weather conditions. Nevertheless, on receiving the order from X. *Fliegerkorps* to return, all but one unit, *Hptm.* Wagner's KGrzbV 103, abandoned the mission. Unlike the rest of the formation, *Hptm.* Wagner's unit was under the direct command of the *Lufttransportchef* (Transport Commander) *Land*, and as his orders did not come from that source, he suspected the signal was an enemy ruse. He therefore continued as planned, and although KGrzbV 103 forced a landing at Oslo-Fornebu, it sustained heavy losses in the face of heavy enemy ground fire.

Among the casualties was *Hptm.* Wagner himself who, approaching in the first machine, was killed by a burst of machine gun fire as his aircraft came in to land. *Hptm.* Ingenoven then took command and organised the Army troops into securing the field. Once this had been achieved, he directed the landing of the rest of his *Gruppe*, the pilots skilfully setting down their machines amidst the wrecked and burning

German infantry preparing to board a Ju 52/3m during the invasion of Norway.

German and Norwegian aircraft which littered the field. Once X. *Fliegerkorps* learned that German aircraft were landing and taking off from Fornebu, the 3. *Bataillon* of the 324. *Infanterie Regiment* was ordered to continue to Oslo instead of landing at Stavanger as planned, in order to make up for the missing paratroops. Since weather conditions over the sea had subsequently improved, the paratroops, who had returned to Aalborg, were also ordered back to the area and landed at Oslo in aircraft from II./KGzbV 1.

By the afternoon of 9 April, the transport aircraft were being used to airlift supplies, one of the first routine tasks being the organisation of a fuel supply to I./ZG 76 at Oslo and Stavanger. Later, large-scale supplies were organised to Stavanger, Bergen and Trondheim, and troop transports were flown to Oslo from Germany.

Only two Junkers G 38s were completed. The first was destroyed in a crash in 1936, but the second, with the Stammkennzeichen GF+GG, was pressed into service during the occupation of Norway when it was assigned to KGrzbV 107 and used to transport military supplies to Oslo-Fornebu. The machine was later assigned to I./KGzbV 172 , and with parts of its tail painted yellow, as seen here, later took part in the Balkans campaign. It was eventually destroyed at Athens-Tatoi airfield during an RAF attack on 17 May 1941.

After the first day, all units except II./KGzbV 1, which was to be at the disposal of X. *Fliegerkorps* for the whole of the operation, came under the control of the newly-created *Lufttransportchef, Land*. Thus I./KGzbV 1, now based at Gardelegen, III./KGzbV 1 at Burg and IV./KGzbV 1 at Braunschweig-Broitzem were assigned to *Lufttransportchef Land* for the task of transporting reinforcements and supplies as necessary to the Danish airfields at Aalborg West and Aalborg East and to Oslo-Fornebu and Stavanger in Norway. In addition, the following *Gruppen* were freshly allocated from training and, on paper, made up KGzbV 2 [2]:

KGrzbV 101	based at Husum	Ju 53/3m
KGrzbV 102	based at Husum	Ju 53/3m
KGrzbV 103	based at Schleswig	Ju 53/3m
KGrzbV 104	based at Stade	Ju 53/3m
KGrzbV 105	based at Kiel/Holtenau	Ju 53/3m
KGrzbV 106	based at Ütersen	Ju 53/3m
KGrzbV 107	based at Hamburg/Fuhlsbüttel	Ju 53/3m, Fw 200, Ju 90, G 38
KGrzbV 108	based at Norderney	He 59, Do 26, Bv 138, Bv 139

A small and extemporary parachute operation took place on 14 April when, on direct orders from Hitler, some 160 paratroops from a reinforced company of 1. *Fallschirmjäger Regiment* at Oslo were dropped by 15 aircraft from II./KGzbV 1 at Douvre and at Dombås, the latter an important rail centre in the Gudbrandsdal Valley. Weather conditions were exceedingly unfavourable, and the aircraft, flying in loose formation, had to circle over the target area for some time before finding a suitable drop zone. Nearly all of them ran into heavy ground fire and one aircraft, complete with crew and paratroops, was destroyed over the target area. Another four machines were so badly damaged that they crash-landed during their return flight. Moreover, strong crosswinds caused heavy casualties among the paratroops, but they succeeded in preventing a link-up between the British at Aandalsnes and the Norwegians to the south. The foul weather also prevented the delivery of ammunition, food and other supplies to the paratroops, but they held the vital communication point for nearly a week.

After the first ten days, the need for air transport declined and units were withdrawn to Germany to refit. However, in May, *General* Eduard Dietl, commanding the 3. *Gebirgsjäger Division*, was cut off at Narvik. The aircraft of KGrzbV 107 and KGrzbV 108 were fitted with long-range tanks and ordered to fly in one thousand replacement troops and supplies. The troops consisted of paratroops from a battalion of Parachute Rifle Regiment 2 which was picked up in Holland [3] to jump again at Narvik, and mountain infantry which volunteered to drop by parachute despite the fact that they lacked formal parachute training.

Munitions and all manner of supplies, including delicate equipment landed by seaplane in a fjord and large numbers of boots for shipwrecked naval personnel, were flown in. In order to provide the Narvik force with heavy weapons, KGzbV 102 under *Oberst* Baur de Betasz was ordered to fly in a battery of

---

2. KGzbV 2 was created in August 1938 but consisted only of a Stab under which further units from the training schools and other establishments would be raised as required for special operations. This process of creating units as and when required was known as the Kommandoweg, or Command Method.
3. The German invasion of Holland, Belgium and Luxembourg had meanwhile begun on 10 May 1940.

75 mm mountain guns and ammunition. Carrying only enough fuel to reach a frozen lake near Bardufoss, which was too short to allow the aircraft to take off again, the aircraft were abandoned and sank during the spring thaw. The crews of these aircraft joined the ground troops and were deployed on the ground.

Once the Norwegian campaign was over, the 617 transport aircraft which had taken part had flown a total of 3,018 sorties, during which 2,376 tons of supplies and 29,280 personnel were flown in for the loss of a third of their number.

ABOVE AND ABOVE RIGHT: Ju 52/3ms at Narvik in April 1940.

The First Glider Operation

In November 1939, the *Luftwaffe* had envisaged the necessity for organising an airborne operation to capture the fortress of Eben Emael, then reputed to be the strongest of its kind, and prevent the demolition of three vital bridges over the Albert Canal when Germany invaded Belgium. With this in mind, *Hptm.* Walter Koch of the *Fallschirmjäger* was ordered to form and train a special *Sturm Abteilung*, or Storm Battalion, with four detachments code-named 'Steel', 'Concrete', 'Iron' and 'Granite'. Group 'Granite', with 11 gliders and 85 men under the command of *Lt.* Rudolf Witzig, was to seize the fortress of Eben Emael, while the bridges at Veldwezelt, Vroenhoven and Kannes were the objectives assigned respectively to 'Steel', under the command of *Oblt.* Altmann, 'Concrete', under the command of *Lt.* Schacht, and 'Iron', under the command of *Lt.* Schächter.

On 9 May 1940, the paratroops were taken from their base at Hildesheim to airfields at Cologne-Ostheim, Cologne-Butzweilerhof and Münster-Loddenheide, where they embarked with their equipment into DFS 230 gliders from KGrzbV 5. At 04.30 hrs the next morning, each glider was towed into the air by a Ju 52/3m from III./KGzbV 1. The first objectives were the bridges at Veldwezelt ('Steel') and Vroenhoven ('Concrete'). In order to camouflage operations, some 120 dummies were dropped by parachute to confuse the defenders and initial losses were very light. The two objectives were taken within 15 minutes, but a heavy machine gun group dropped 45 minutes later into the same area became separated from their weapons and equipment containers and suffered heavy losses before they were relieved by forward elements of the German Army.

RIGHT: The Western campaign opened on 10 May 1940 with a bold airborne assault to seize the fort at Eben Emael and three important bridges over the Albert Canal. The airborne force, known as Sturmabteilung Koch, after its commander, Hptm. Walter Koch, comprised 11 officers and 427 men of I./FJR1 and the Pioneer Kompanie of II./FJR1 under Lt. Rudolf Witzig. The force was divided into four assault groups known as 'Granite', 'Steel', 'Concrete' and 'Iron'. Assault Group 'Granite' was led by Lt. Witzig and carried out the assault on the fort. This photograph shows one of the fort's retractable armoured rotating cupolas, while two of the DFS 230 gliders used in the operation can be seen in the background. Hollow charges were employed to destroy the armoured cupolas.

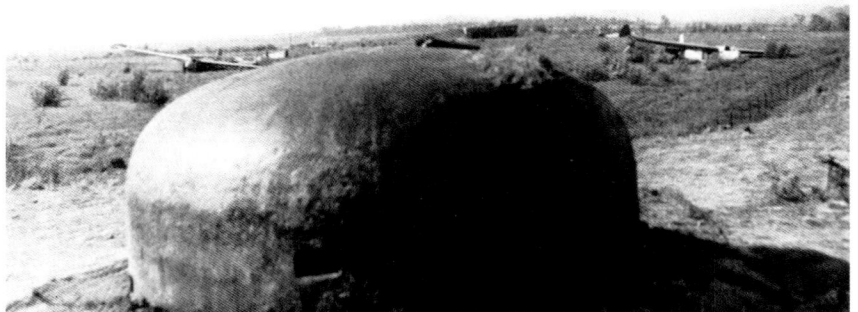

The detachments landed by glider at the bridge at Kannes ('Iron') and Eben Emael ('Granite') were not so immediately successful. At Kannes, the bridge was blown up at the very moment 'Iron' detachment's gliders approached to land, slowing down the Army units advancing to relieve 'Granite' which had, in the meantime, encountered stiff resistance at Eben Emael. Nevertheless, by the morning of 11 May, *Lt.* Witzig was able to hand over the captured installations to German engineers who had reached the fort during the night.

ABOVE: Lt. Rudolf Witzig, the commander of the glider force assigned to attack the Eben Emael fortifications. During the flight to the target, the Ju 52/3m towing his glider was obliged to manoeuvre to avoid another aircraft and the tow rope parted. Witzig's glider landed near Cologne and only after another Ju 52/3m, complete with a new tow rope, was made available did he arrive at the fort. He landed some two hours after the main glider force had begun the assault on the casements and cupolas.

Air transport proper, as distinct from the Koch operation, began at 04.00 hrs on 10 May when Ju 52/3ms took off from airfields in north-west Germany and headed for the Dutch border in what was to be the *Luftwaffe's* first large-scale operation by parachute and air-landing forces. At 05.30 hrs the first wave of aircraft, comprising *Obstlt.* Fritz Morzik's KGzbV 1, crossed the border in a broad front formation, dropping down to hedge-hopping height for the rest of the journey to their objectives. These were the airfields at Waalhaven, near Rotterdam; the airfields at Valkenburg, Ypenburg and Ockenburg near The Hague; and the bridges at Moerdijk and Dordrecht.

The Moerdijk viaduct was taken undamaged, but at Dordrecht the Dutch recaptured the bridge and the paratroops became locked in bitter fighting for the town. At Waalhaven, *Hptm.* Zeidler's III./KGzbV 1 dropped paratroops just outside the airfield but they were still battling with the enemy when KGzbV 2, under the command of *Oberst* Gerhard Conrad, arrived to land elements of 22. *Infanterie Division*. However, the enemy had mined the landing field and several aircraft were destroyed as they touched down. Eventually, through the combined efforts of the airborne troops and paratroops, the airfield was secured.

Unfortunately, paratroops which should have jumped onto the three airfields near The Hague were dropped in the wrong places and the Ju 52/3ms air-landing other elements of 22. *Infanterie Division* were compelled to land at airfields still in enemy hands. Moreover, the airfields had been strongly fortified and rigged with barriers. At Valkenburg, the first *Gruppe* of Ju 52/3ms sunk into the soft ground on the airfield and bogged down, which prevented the second wave from landing and enabled the Dutch to recapture the airfield. At Ypenburg, the anti-aircraft fire was so fierce that 11 of the first 13 Ju 52/3ms which attempted to land were shot down or crash-landed onto the airfield, again preventing further aircraft from landing. Soon the sky over the three airfields was filled with aircraft from KGrzbV 9, KGrzbV 11, KGrzbV 12 and KGrzbV 172, all trying to find a place to touch down. In desperation, they eventually landed where they could: in open countryside; on beaches where they sank up to their axles in sand and shingle; or on roads and highways where they suffered equally heavy losses. Particularly badly affected was *Obstlt.* Johannes Janzen's KGrzbV 9 which, on this single day of operations, reported a total of 39 Ju 52/3ms destroyed or damaged beyond repair, thus eliminating it from further participation in the campaign.

Later in the day, successive waves of aircraft left Münster-Loddenheide, Lippspringe and Gütersloh bound for Rotterdam, Nijmegen and Delft. The aircraft carried Army troops, bicycles, motorcycle combinations, munitions etc, while signals Ju 52/3ms brought complete airfield W/T installations with them. Aircraft were coming in everywhere; those which could take off again after delivering their cargoes made up to three sorties on this first day.

Twelve He 59 seaplanes landed on the River Maas at Rotterdam and brought in a company of 22. *Infanterie Division* to capture and hold bridges until relieved by the motorised advance. Some 500

ABOVE: Officers of the assault force after being decorated with the Ritterkreuz by Adolf Hitler. Hptm. Walter Koch, who was placed in overall charge of the operation, is in the front row, second from left, while Lt. Rudolf Witzig is in the back row, first left.

ABOVE: During the invasion of Denmark and Norway, KGzbV 2 controlled a number of He 59s of KGrzbV 108. These seaplanes were also employed in Holland when twelve He 59s landed on the River Maas at Rotterdam and brought in troops to capture and hold bridges.

other personnel, believed to be from 7. *Infanterie Division* and apparently organised spontaneously, were landed by various light aircraft in the area between Bastogne and Libement. Here, aided by dummy paratroops to confuse the enemy, they disrupted communications and acted as sabotage parties.

In the first day of the operation against Holland and Belgium, some 350 troops were landed by glider, some 4,000 paratroops were dropped and the whole of 22. *Infanterie Division* and 72. *Infanterie Regiment* landed by air. On the second day, ammunition was dropped to paratroops at Dordrecht, and on the third day still further troops were landed at Rotterdam-Waalhaven.

The rapid advance of the invading Army made further air-supply operations unnecessary, and after 15 May all Ju 52/3m units except KGzbV 1 were withdrawn and returned to the training units. KGzbV 1 was allocated to *Luftflotten* 2 and 3 for the transport of supplies to France and Belgium, and although these operations were conducted on a makeshift basis, several important missions were flown. Thus, on one occasion, transport aircraft carried bombs to a Stuka unit stationed on an advance airfield in France. The bombs were unloaded and immediately fitted to the waiting Ju 87s which released them in support of attacking forces a short distance from the field. For the first time, the evacuation of wounded was added to the duties of the air transport forces and casualties were flown to the rear in aircraft making their return trips.

In August 1940, all KGzbV units were ordered to be in readiness for 'Sea Lion', the invasion of England, but were eventually stood down and most were returned to the C-*Schulen* when the operation was postponed indefinitely on 17 September. In the transport and airborne forces, it was widely believed that the cancellation resulted from the fact that the British had erected so many obstacles in the selected landing areas that an airborne operation could not succeed. In fact, Hitler had decided that Britain would be subjugated by economic warfare and placed his reliance on the U-boat force to starve the British into submission while he turned his attention eastwards towards Soviet Russia.

BELOW: Tasks were assigned to transport Gruppen by the Luftflotte or Fliegerkorps to which they were subordinated. These included supplying Luftwaffe units with all the more urgent requirements of an operational unit including bombs, ammunition, petrol, oil and oxygen, kitchens, stores, rations and drinking water. Moreover, when an operational Gruppe moved to a new base, aircraft of the transport Gruppen were often assigned the task of transporting them. Here, Ju 52s are taking off after moving ground personnel and equipment to a forward airfield.

RIGHT: During the attack on Holland, the 22. Infanterie Division was to have been landed on three airfields near The Hague, but the Dutch defenders had strongly fortified them and prepared obstacles to prevent landings. Paratroops who should have secured the airfields were dropped in the wrong places and many Ju 52/3ms were destroyed as they touched down.

ABOVE: Oberstleutnant Gustav Wilke was awarded the Ritterkreuz on 24 May 1940 while Kommandeur of KGrzbV 12. This Gruppe existed only for some four months and, after being employed during the invasion of Holland, was disbanded in June 1940.

ABOVE: Many Ju 52/3m were destroyed or badly damaged by Dutch opposition while landing at Rotterdam-Waalhaven airport in May 1940 to land troops who then came under fire as they left the machines. This scene shows transports of II./KGzbV 1, with smoke from the fighting still rising in the distance.

RIGHT: Three damaged Ju 52/3ms with another burning in the background, after receiving a direct hit from a Dutch mortar bomb.

ABOVE: An aerial view of Rotterdam-Waalhaven taken as Ju 52/3ms of the first wave take off to pick up more troops. Note the number of closely-packed aircraft still on the ground.

LEFT AND BELOW: The remains of several Ju 52/3ms which crash-landed in Holland. The machine shown (*LEFT*) was photographed near the airport at Ypenburg.

1939-1943

ABOVE: There were many different types of Ju 52/3m with interiors designed or adapted for a particular purpose. Some were ambulance aircraft, as shown here, and could carry 12 stretcher cases or up to 24 sitting cases, although the use of such specialised aircraft at the front was rare; as a rule normal Ju 52/3m transports were employed for casualty evacuation. Other machines were fitted out as signals aircraft and were equipped with special W/T equipment and a temporary aerial to act as advanced W/T ground stations until a normal signals station could be established. Some Ju 52/3ms were fitted out as flying workshops, complete with small lathes and their own power units, and there were also Ju 52/3ms fitted out as refrigerators for the transport of meat and other perishable rations to tropical areas.

RIGHT: The interior of an ambulance Ju 52/3m looking aft and showing the supports for the stretchers.

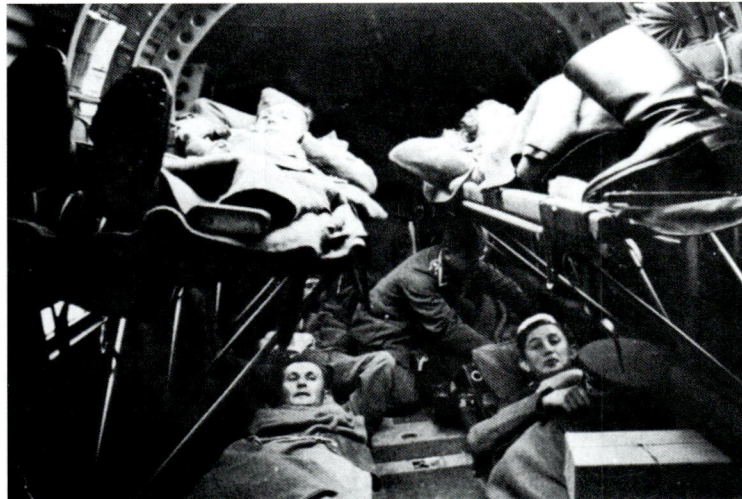

BELOW: This Ju 52/3m belonged to Transportstaffel/V. Fliegerkorps which, from July 1940 to June 1941, was based at Villacoublay. It is believed that this machine was used by Generalfeldmarschall Robert Ritter von Greim, then commander of V. Fliegerkorps. The white stripes on the engine cowlings have been observed on other Ju 52/3ms, but the presentation of the unit code, G5+HH, with the 5 and first H in white is probably unique.

The Invasion of the Balkans

Once German forces were firmly established in France, the Low Countries, Denmark and Norway, there was little or no need for air supply in the West. In central and southern Europe, however, Italy's inability to establish superiority in the Mediterranean resulted in ever-increasing German commitments and the employment of air transport.

The first German air transport operations in the Mediterranean arose from the Italian campaign in Greece, which had opened in October 1940. Despite initial successes, Italian troops were soon forced to retreat into the Albanian mountains, and although the Italians possessed perfectly suitable SM.81s, they requested their German allies to come to their aid. At the end of November 1940, the more experienced III./KGzbV 1 was therefore detached from the rest of the *Geschwader* and, on 8 and 9 December, its 53 Ju 52/3ms flew to Foggia where they were to operate under a *Luftwaffe* Staff in Italy. The *Gruppe*, known during the 53-day operation as '*Transgerma*', flew 4,028 sorties and carried 30,000 personnel, some of whom were trained and equipped for winter warfare, plus 4,700 tons of winter clothing, ammunition, weapons and other equipment to the Albanian airfield of Tirana. No enemy action was encountered during these sorties or during the loading and unloading activities. Indeed, the whole mission was carried out under almost peacetime conditions with no aircraft losses. On their return trips, the *Gruppe's* aircraft transported about 10,000 wounded and sick back to Italy.

Meanwhile, X. *Fliegerkorps*, which now controlled all operational flying units in the Mediterranean theatre, had moved to Sicily, and on 12 December KGrzbV 9 with 30 Ju 52/3ms moved to Reggio to transport *Luftwaffe* units from Italy to Sicily. However, as it became clear that Germany would have to come to Italy's further assistance in the Balkans and Mediterranean, it was realised that a very much larger air transport force would be required. Therefore, in January, KGzbV 40, KGzbV 50 and KGzbV 60 were formed, followed by KGrzbV 101, KGrzbV 104 and KGrzbV 105 which were called up in March. Some of these units were used to transport the ground personnel and equipment of *Luftwaffe* flying units in preparation for the Balkans campaign, while I. and II./KGzbV 1 moved parts of the former 22. *Infanterie Division*, by now renamed 22. *(Luftlande) Infanterie Division*, to Rumania and Bulgaria.

During April, air transport in the western Mediterranean was allocated to KGrzbV 9, KGrzbV 104 and III./KGzbV 1 which between them carried men and material between Italy and North Africa, and Italy and the Balkans. Towards the end of April, these *Gruppen* were reinforced by the temporary

Despite the code G6 being used by a number of units, the Staffel letter 'T' identifies this Ju 52/3m as having belonged to KGrzbV 104 which was formed in March 1940 and operated until December 1942 when it was redesignated to form parts of KGzbV 323. The photograph was taken at Zeltweg in Austria in January 1941, and as this was the base of Flugzeugführerschule (C) 11, it suggests that this aircraft had been called up from the school for operational duties with KGrzbV 104 in the Balkans and Mediterranean. That the operational markings G6+HT were allowed to remain is not unusual as they were regarded in the schools as battle honours. This particular machine appears to have been camouflaged 71 overall but with the lower part of the centre engine cowling in a light colour, possibly white, which was the Staffel colour of 1./KGrzbV 104, or yellow. Note also that the Hakenkreuz on the tail is in the early position and the individual aircraft letter H was repeated under the wingtips. The emblem on the nose cannot be identified. KGrzbV 104 was one of the units called up in March for the Balkans campaign.

The Corinth Canal, an important waterway which allowed passage from the Gulf of Corinth to the Aegean Sea. The Corinth bridge was the scene of a German parachute assault on 26 April 1941.

transfer of II./KGzbV 1 and KGzbV 172 [4] from the Balkans to Foggia. Large formations of sometimes up to 30 Ju 52/3ms were organised to transport petrol and troops to Benina, Derna and further eastwards, the whole operation being apparently staffed by the original 'Transgerma'.

On 6 April 1941, Germany invaded Yugoslavia and Greece under the code name 'Marita'. In Yugoslavia, German forces were in Salonika by 9 April and by the 13th had entered Belgrade. In order to prevent the British from forming a strong defence in Greece it was important that German forces should secure the Corinth Canal which separates the Peloponnese from the mainland and which formed a natural barrier. Although the canal was easily defended even by a small force, the Germans were aware that the British would make every effort to strengthen the defences in depth. It was therefore necessary to occupy the Isthmus of Corinth at the earliest moment. Moreover, a single bridge across the canal had to be captured intact to allow German ground forces to continue their advance, and the only means available to accomplish such an objective was a surprise attack by airborne troops.

Overall control of the operation was assigned to *Generalleutnant* Wilhelm Süssmann, then the commander of 7. *Fliegerdivision*. *Generalleutnant* Süssmann was ordered to occupy the Isthmus on both sides of the canal with the whole of the reinforced *Fallschirmjäger Regiment* 2 comprising some 2,220 officers and men. They were then to subdue enemy resistance and defend the occupied area until relieved by troops advancing from the mainland. The parachute troops were to be dropped over the target area by the Ju 52/3ms of KGzbV 2 under *Oberst* Rüdiger von Heyking, and this unit was to be responsible for any air-supply operations which might prove necessary. Although planned at very short notice, the timing of the operation was critical since any delay could result in complete failure.

In order to ensure absolute secrecy, the undertaking was to be carried out from the airfield at Plovdiv, in Bulgaria, although on the night immediately preceding the operation, the Ju 52/3ms were to make an intermediate landing at Larissa in Greece to refuel. The operation was to be carried out on 26 April, and on the 25th, the transport units and the paratroops were ordered to stand by for immediate action. Briefing for the accomplishment of the mission was very thorough, with the move to Larissa, the refuelling operation and the take-off line-up being discussed in detail. At the same time, the paratroops were distributed among the transport units and assigned their specific aircraft. Although under the overall control of KGzbV 2, the 270 aircraft to be used in the operation were drawn from I. and II./KGzbV 1, KGrzbV 60, KGrzbV 102 and I./LLG 1. The latter also included ten DFS 230 gliders, while in reserve were KGrzbV 101, KGrzbV 172 and a bomber *Gruppe*.

In order to avoid alerting the enemy by concentrating too many aircraft at Larissa in daylight, the first formations were timed to arrive there at dusk. However, delays in the loading of the aircraft resulted in the original take-off schedule being disrupted and, instead of all aircraft of a unit taking off together, they set out singly or in small groups. The result was that aircraft from the same unit were unable to stay together and became mixed up with those from different units.

Although all machines arrived at Larissa before nightfall, the effects of the disorganised take-off had to be rectified in order to avoid delays in the take-off next morning. During the night, the units were therefore reassembled and each aircraft was lined up in its proper position. Additionally, because the necessary refuelling arrangements had been overlooked in the planning of the operation, as each one of the 270 aircraft landed, it had to taxi to a central fuel depot where it was refuelled from a drum by means of a handpump. Once refuelled, the aircraft taxied to its allocated position where the crew and the parachute troops assigned to it were ordered to remain with it. Due to the favourable weather, this presented no difficulty and it was later estimated that the paratroops obtained more rest this way than if they had marched long distances to billeting areas elsewhere. However on account of the number of aircraft involved and the primitive facilities available, refuelling lasted almost until the morning and was only completed as the first unit was lining up for take-off.

Despite the difficulties encountered during the night, the first aircraft took off on schedule while it was still dark, shortly before 05.00 hrs on 26 April. First to leave were three aircraft each towing a freight glider which contained the troops assigned to seizing both ends of the canal bridge and removing any demolition charges presumed to have been placed there. These aircraft left ahead of the

---

4.   In May 1941, Kampfgeschwader zbV 172 was reduced to a single Gruppe, I./KGzbV 172. Although sometimes referred to as KGrzbV 172, or Kampfgruppe zbV 172, this is thought to be incorrect.

main force as towing the gliders slowed them down. An allowance had been made for this, however, and the aircraft following would catch up, thus ensuring that the gliders and the first paratroops would arrive at the target simultaneously at 07.00 hrs. The next transport aircraft left on schedule, taking off in groups of three, each close behind the other, and although subsequent units were delayed, they were able to make up for any lost time during their two hour flight.

Once in the air, the machines flew towards the Corinth Canal in a long stream of three aircraft abreast, each pilot making use of the exhaust flames from the aircraft ahead to assist him in maintaining his course until it grew light. As the aircraft flew at only about 30 metres and there was a slight haze, they reached the target undetected. During the last few minutes of the flight, as each aircraft climbed to 120 metres to release its paratroops, bombers and dive-bombers from VIII. *Fliegerkorps* bombed and strafed the target area to force the defenders to take cover. As the last of the bombs, the gliders and the paratroops all landed simultaneously, the enemy was taken completely by surprise and both ends of the bridge were quickly seized as planned. Unfortunately, however, all the removed demolition charges were still piled on the bridge when, in a bizarre accident, a stray shell from a British anti-aircraft battery landed in the explosives and they detonated immediately. Although the bridge was destroyed, engineers of the *Fallschirmjäger* construction platoon, which had landed in the gliders, were able to construct an emergency bridge alongside the old one and the way was cleared for advancing Army troops to cross from the mainland to the Peloponnese.

Such was the success of the operation that no reinforcements or supplies had to be flown to the troops holding the bridge and their only request was for anti-tank mines. These were landed on an emergency airstrip near the canal by a single Ju 52/3m of KGrzbV 102, and subsequent ground action was such that the transports could be immediately withdrawn.

ABOVE: The Ju 52/3ms of 11./KGzbV 1 had a yellow individual aircraft letter and a yellow background to the Staffel emblem, upon which was a black bear. This aircraft carried the operational code IZ+DV and was photographed early in the Balkans campaign. Note that the Ju 88 and the Ju 87 Bs in the background already have yellow rudders.

ABOVE: The 5. Staffel of KGzbV 1 photographed during a flight from Cottbus to Belgrade in 1941. The aircraft centre left has the Geschwader's Brandenburg eagle emblem superimposed upon a yellow band painted between the cockpit and the engine. Note the different locations of the swastikas on the tails of the two machines shown on the right, the machine nearest the camera having the swastika in the pre-war position.

RIGHT: Ju 52/3ms of II./KGzbV 1 over Belgrade in 1941. Although both aircraft carry KGzbV 1's emblem on the forward fuselage, they appear to be marked only with their individual aircraft letters. Also visible on the machine nearest the camera are the yellow markings adopted during the Balkans campaign, although in this case they have been applied only to the outer engine cowlings and elevators.

On 6 April 1941, German forces attacked Yugoslavia and Greece with forces based in Austria, Hungary, Rumania and Bulgaria. This photograph shows a Ju 52/3m flown by the Staffelkapitän of 15./KGzbV 1 over Bulgaria with the yellow central cowling and rudder applied during the Balkans campaign. During the campaign in Yugoslavia and Greece, most of the transport units attached to XI. Fliegerkorps were engaged in carrying supplies.

Emblem of IV./KGzbV 1

Junkers Ju 52/3m of Stab, 15./KGzbV 1, Balkans, 1941

This Ju 52/3m was finished in the standard 70/71/65 camouflage scheme with the centre engine cowling and rudder painted yellow as a recognition aid during the Balkans campaign. The emblem on the nose was that of IV./KGzbV 1, to which the 15. Staffel belonged, and the machine carried the operational markings IZ+AZ with the individual aircraft letter repeated in yellow on the leading edge of the wing.

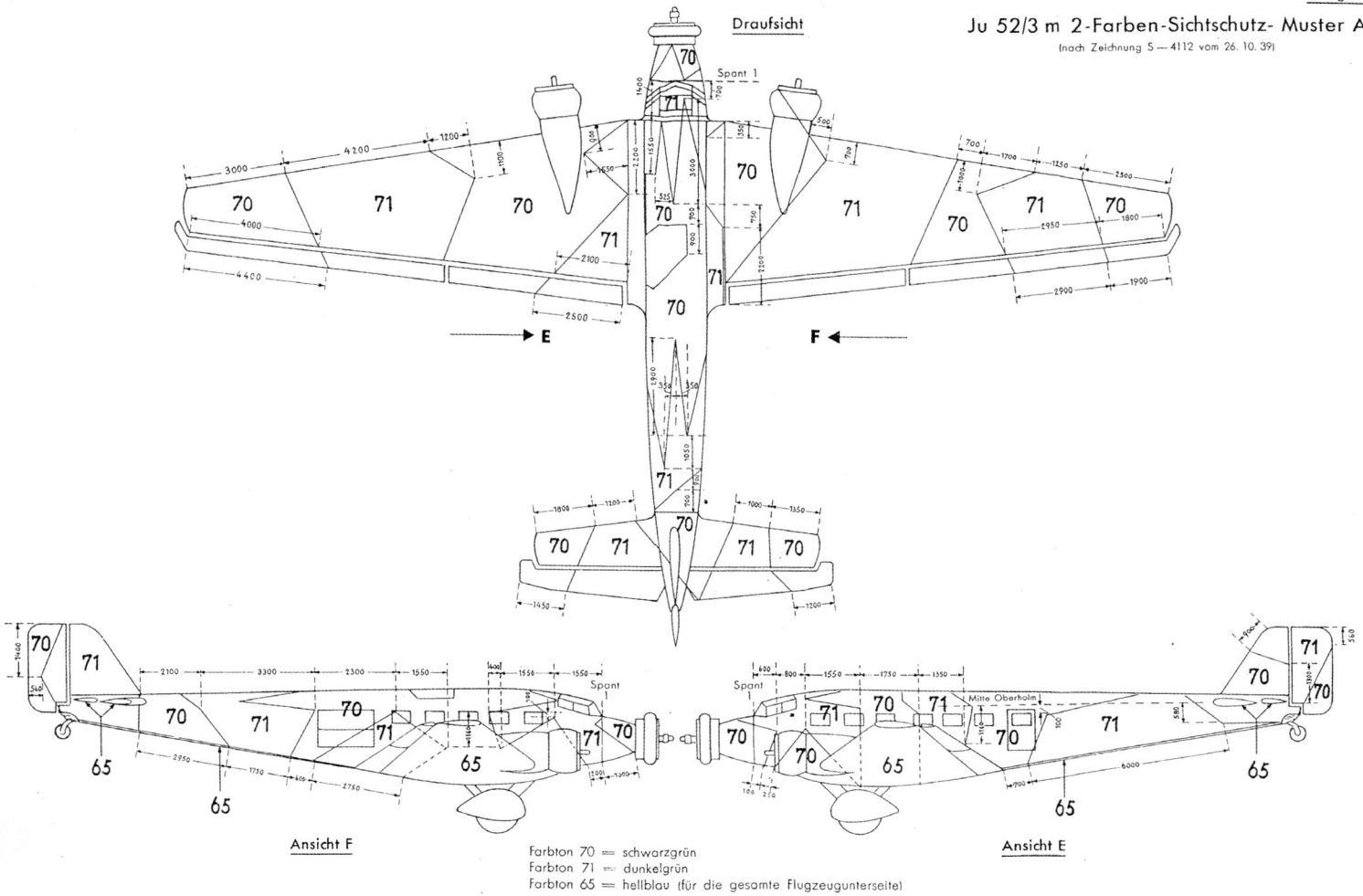

Anlage 7

Ju 52/3 m 2-Farben-Sichtschutz- Muster A
(nach Zeichnung S — 4112 vom 26. 10. 39)

Draufsicht

Ansicht F

Ansicht E

Farbton 70 = schwarzgrün
Farbton 71 = dunkelgrün
Farbton 65 = hellblau (für die gesamte Flugzeugunterseite)

ABOVE: An official drawing showing the approved RLM splinter scheme for the Ju 52/3m, Type A. A mirror image could be applied as Type B.

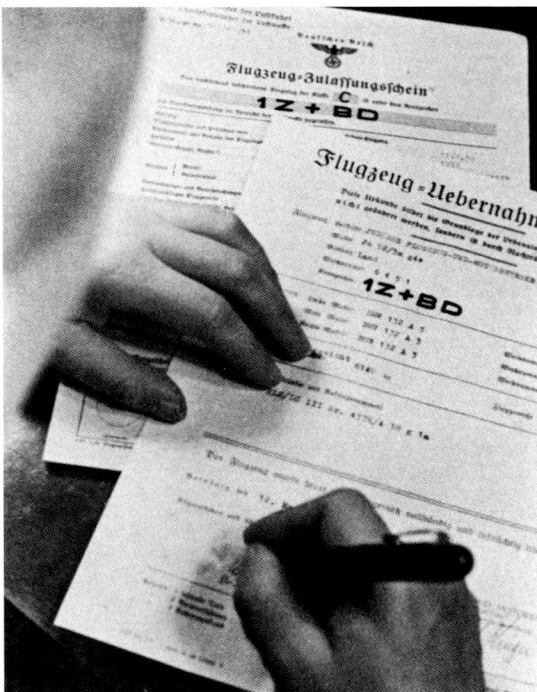

LEFT AND ABOVE: Once the camouflage had been applied and the aircraft delivered to its operational unit, the paperwork had to be completed Trand the operational markings applied. The documents shown (*LEFT*) are the aircraft licence and, being signed, the aircraft acceptance form. Both forms show the operational markings, which were then applied to the fuselage sides (*ABOVE*), in this instance in white. Hand painting was required as the corrugations rendered the use of stencils impractical. Note also the trace of the factory code, just visible inside the letters B and D.

Unternehmen 'Merkur'

Although the strategic importance of the possession of Crete was appreciated by the German High Command, an airborne attack on the island was not a part of the original 'Marita' operation in the Balkans. However, in March 1941, *Generaloberst* Kurt Student, commander of the newly-established XI. *Fliegerkorps*, drew up a plan of attack and presented it to Hitler on 21 April. On 25 April Hitler approved the operation, which was to be code-named *'Merkur'* and was to take place on 17 May. This gave XI. *Fliegerkorps* only 22 days in which to make the necessary preparations for the largest airborne operation yet attempted. The plan was to attack in two waves, the first in the morning to capture Maleme airport, Canea town and the port at Suda. The second wave, mounted later in the day, was intended to take the towns and airports of Rethymnon and Heraklion. The success of the operation depended on the early capture of an airport so that reinforcements could be flown in.

Some 500 aircraft comprising ten *Transport Gruppen*, two of which were equipped for glider-towing, were made available but after flying almost continuous daily supply operations in the Balkans they were urgently in need of maintenance. On 12 May, therefore, the whole transport fleet flew to maintenance centres in Germany, Austria and Czechoslovakia where they were rapidly overhauled and many were fitted with new engines. Meanwhile, DFS 230 gliders were brought up by rail and by 19 May, the transport aircraft and gliders were ready on six airfields in Greece. However, more problems had yet to be overcome before the operation could be launched.

The greatest of these difficulties concerned the transport of supplies into the Athens area. This was due in part to the destruction of rail communications during the war in Yugoslavia, but also to the advancement of Army units moving up for the forthcoming invasion of Russia which clogged the roads in Rumania. Consequently, the 22. (*LL) Division*, which was supposed to take part in *'Merkur'*, could not get through and the 5. *Gebirgsjäger*, a mountain division already in Greece, had to take its place. Moreover, the airfields which had to accommodate the ten *Transport Gruppen* lacked refuelling facilities. Each aircraft would require some 2,000 litres for each sortie and XI. *Fliegerkorps* had decided that, at least for the start of the operation, sufficient fuel for each aircraft to make two or three sorties per day was to be stocked at the airfields. The full ten *Gruppen*, comprising over 500 aircraft operating at full strength, would therefore require more than 900,000 litres of fuel for their first sorties alone. Due to the transportation bottlenecks on land, the fuel could only be delivered by sea, but only the airfields at Corinth and Megara were in the vicinity of suitable ports and all fuel delivered to these airfields was in barrels. These barrels were then distributed by road to other airfields, a process that imposed a delay in the attack of two days.

Finally, the airfields themselves were exceedingly dusty. At Topolia, *Oberst* Rüdiger von Heyking, the *Kommodore* of KGzbV 2 controlling KGrzbV 101, KGrzbV 102, KGrzbV 105 and KGrzbV 106, was horrified to discover that an over-enthusiastic Army officer had arranged for the airfield to be ploughed up after its occupation to make it more level. Subsequently, every take-off and landing produced such a cloud of dust that it took 17 minutes before it settled and the field could be used again.

The attack was now postponed until 20 May, yet difficulties were encountered up until the last minute. Refuelling of the Ju 52/3ms by handpump had to be carried out by the troops who were due to go into action the next day, and during the night the wind changed 180 degrees so that all the aircraft already lined up at one end of each of the airfields had to be manhandled to the opposite end. Water wagons sprayed the airfields in an attempt to lay the dust. The troops themselves were informed of their targets only on the day before the operation. Arriving at their airfields after dark, they were greeted by the roar of the transports and clouds of dust as the aircraft engines were tested.

The airborne attack on Crete began on 20 May when I./LLG 1 and parts of KGzbV 1 towed 53 DFS 230 gliders off from Tanagra, landing the HQ Staff and parts of the I. *Bataillon* of the *Luftlande Sturmregiment* west and south of Maleme airfield at 07.15 hrs. KGzbV 2 followed soon afterwards, dropping further elements of I. *Bataillon* near the same objective. However, the Germans had miscalculated the strength of the Commonwealth garrison on Crete which proved to be between three to five times stronger than they had estimated. Moreover, the defenders had been passed intelligence material that

General Kurt Student, former commander of 7. Fliegerdivision and, in 1941, commander of XI. Fliegerkorps, the Luftwaffe's parachute command. Student planned the airborne assault on Crete and, after persuading Göring and Hitler of the feasibility of the plan, was given just 22 days in which to make the necessary preparations.

included the complete German battle plan, and in anticipation of the airborne assault had prepared the defence of the areas of operations with the greatest care and with every possible means. German losses were, therefore, high from the start. A *Gruppe* bringing in a *Fallschirmjäger Bataillon* seriously overestimated wind drift and carried the men so far inland that they jumped over New Zealander gun positions and landed in unsuitable terrain. Many *Fallschirmjäger* were killed while still in the air or were injured on landing; all this *Bataillon's* officers lay dead or wounded, and the survivors were pinned down by Allied defensive fire. Glider and parachute troops landing at Canea were equally unsuccessful and by midday, none of the prime objectives had been secured. A feature of German parachute operations was that the *Fallschirmjäger* jumped without weapons and only little ammunition, these being dropped separately in containers. However, many paratroops were separated from their weapons while others soon ran low of ammunition.

Meanwhile, the transport aircraft had returned to their airfields, but the dust raised by the first formations to land compelled the remainder to circle for up to two hours before they could touch down. Many aircraft crashed into each other in the dust and confusion, with the result that losses on the ground far exceeded those from anti-aircraft fire over Crete and the timing for the second wave, which was to attack the airfields at Rethymnon and Heraklion, went to pieces. Refuelling difficulties, repairs, the clearing of wrecks from the airfields, the heat and the dust so delayed matters that most of the aircraft which should have

FAR LEFT:
A paratrooper on Crete retrieving a weapons container.

LEFT: Two Ju 52/3ms of IV./KGzbV 1, almost certainly photographed in June 1941. At that time the Gruppe was employed in Greece, Bulgaria and Yugoslavia to bring up supplies and ammunition for the troops who had recently landed on Crete. The aircraft in the background, IZ+BX, belonged to 13. Staffel and has the standard 70/71 uppersurfaces, whereas the aircraft in the foreground, which belonged to the Gruppe Stab, has a three-tone uppersurface scheme. Note the use of a light grey, and that one of the two greens on this Stab machine, 1Z+BF, is much lighter than 71. This was possibly a hybrid scheme in which the pre-war colours 62 and 63 were used with 71.

made up the second wave were still on the ground when the bombers, fighters and dive-bombers of VIII. *Fliegerkorps* began their softening up bombardment. Eventually, small formations of transports made their way to Crete individually, but the planned mass attack took place only as a number of spasmodic drops over a period of three and a half hours. Not surprisingly, the paratroops again suffered heavy losses as the defenders had largely recovered from the effects of the bombing. Many paratroops were killed before they reached the ground and whole companies were annihilated.

After the first day's operations, the serviceability of the Ju 52/3ms fell alarmingly as a consequence of heavy damage by gunfire, collisions and crash landings. On 21 May, and with the knowledge that losses would be high, XI. *Fliegerkorps* decided to force a landing at Maleme airfield which then became the main point of attack. The first Ju 52/3ms to land during the late afternoon contained the 100. *Gebirgsjäger Regiment* which was unloaded in the face of very heavy artillery and machine gun fire. One out of every three Ju 52/3ms that touched down was destroyed or damaged, but still they came, taxiing among the bursting shells to bring in more and more troops. By the evening, the whole of 100. *Geb. Rgt.* had been landed, together with additional units of 5. *Gebirgs Division*. This air-landed force, together with the paratroops who had in the meanwhile been supplied with fresh ammunition by other Ju 52/3m units, finally turned the day in the Germans' favour, although Maleme was not entirely in German hands until the following day.

Supply dropping was undertaken on a large scale from the second day and heavy artillery pieces, which had to be dismantled before they could be loaded, were transported to Crete by air. At the same time, VIII. *Fliegerkorps*, bombing in support of the operation, had the services of IV./KGzbV 1 and KGrzbV 50 working in Greece and further back to bring up its supplies and ammunition. From 23 May, a *Fallschirmjäger Feldlazarett* (Field Hospital) which had landed on Maleme airfield was in full operation and the wounded were flown back to Athens, or to Bucharest by aircraft flying there to fetch ammunition and other stores. Another task for the air transport units at this time was to organise and supply large quantities of drinking water to meet a constant demand from the troops on Crete.

Eventually, the Germans occupied Canea and Suda and pressed on to relieve their forces at Rethymnon and Heraklion. By the time the operation ended on 1 June, the air transport units had transported some 24,000 men into Crete including about 13,000 paratroops and 9,000 Mountain Troops, but losses in men, gliders and aircraft were very heavy indeed: 271 Ju 52/3ms were destroyed or had to be written off as a consequence of crashes, landing on rough ground, on beaches, or in the face of intense gunfire. KGrzbV 40 and KGrzbV 60 were disbanded and the aircraft passed on to the hardest hit units to restore their strength. Of more than 6,500 troops killed or wounded during the ten-day campaign, 5,140 were from the *Fallschirmjäger*. Never again would such a large-scale parachute operation be attempted and apart from a few isolated jumps, the *Fallschirmjäger* force was confined to a ground role for the rest of the war.

BELOW: Fallschirmjäger loading their equipment into a very weathered Ju 52/3m of KGzbV 2 in readiness for Unternehmen 'Merkur'.

ABOVE: The airborne assault on Crete began with DFS 230 gliders landing 15 minutes ahead of the parachute force. Here, Ju 52/3ms are seen towing the gliders off from Eleusis airfield in Greece.

BELOW: Before 'Merkur', an airfield inspection officer decided that the grass landing ground at Topola was too uneven for transport operations and he had it ploughed up to make it level. In the hot, dry weather, this resulted in the slipstreams from dozens of aircraft raising huge clouds of dust, as shown by this Ju 52/3m transport as it begins its take-off run. On the first day of 'Merkur', the dust on overcrowded airfields lingered in the air for so long that it prevented following aircraft from taking off and further problems were caused by shot-up aircraft crashing and blocking the runways. The schedule which called for returning Ju 52/3ms to rapidly reload and be ready for another mission was therefore delayed and the planned operational timetable for the second wave of troop landings was seriously disrupted with grave consequences.

ABOVE: The jump procedure of the Fallschirmjäger differed in a number of respects from those later employed by the airborne forces raised in Britain and the US. As is shown in this photograph, each Fallschirmjäger leaped from the exit door with both arms outstretched and fell past the horizontal before the static line opened the parachute. The troops therefore landed unarmed and relied on accurate drops of weapons and ammunition containers. Moreover, and unlike the later British X-type parachute that allowed the descent to be controlled by means of pulling on the risers, the Fallschirmjäger was attached to his parachute at a point situated near his shoulder blades. Manoeuvring while in the air or controlling the rate of descent was therefore impossible, and on landing, the Fallschirmjäger was thrown forward onto his hands and knees with a consequently high casualty rate from injuries to wrists and knees.

RIGHT: Although almost certainly taken during an exercise, this view provides a good impression of a parachute drop in progress. Three aircraft each carrying a maximum load could release twice the number of Fallschirmjäger shown here.

Gebirgsjäger preparing to board Ju 52/3ms during the latter part of the invasion of Crete. The aircraft nearest the camera, coded 1Z+LL, belonged to 3./KGzbV 1 and carried the emblem of I./KGzbV 1 on the nose. A feature of aircraft operating in the Balkans was that the engine cowlings, rudders and elevators were painted yellow.

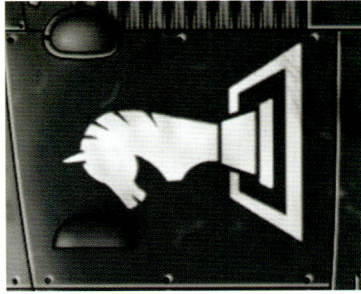

Emblem of I./KGzbV 1

Junkers Ju 52/3m of 3./KGzbV 1, Greece, May 1941
This machine was finished in the standard 70/71/65 camouflage scheme with large areas of yellow. Note, however, the difference between the tone of the yellow in the aircraft letter L and that on the rudder, the latter appearing darker due to the method of application.

1939-1943

RIGHT: Mountain troops inspecting the Ju 52/3m which will carry them to Crete. The 100. Gebirgs. Rgt. began landing on 21 May on Crete in the afternoon of the second day of the assault, followed the next day by 85. Gebirgs.Rgt. and 95. Gebirgs-Pioneer Bataillon.

BELOW: A trail of wreckage and the burning remains of a Ju 52/3m shot down during the Crete operation.

ABOVE AND TOP: The operation to capture Crete in May 1941 met with stiff British and Commonwealth resistance and very nearly ended in a German defeat. Only when the decision was taken to land reinforcements, supplies and heavy weapons on Maleme airfield, which was still under enemy fire, was the balance tipped in favour of the attackers although losses were severe. Evidence of XI. Fliegerkorps' determination to secure the airfield, regardless of losses, is shown in these views of wrecked Ju 52/3ms at Maleme.

RIGHT: An aircraft graveyard at Maleme. Note the completely burned-out remains in the foreground.

1939-1943

RIGHT: Photographed a few days after the main assault, this Ju 52/3m over Greece is believed to be flying mountain troops to Crete. As with all German aircraft types serving in the Balkans, large areas of the aircraft have been painted yellow, but this machine is unusual in that the yellow nose cowling has been enhanced by a yellow band around the forward fuselage.

LEFT: Although lacking the yellow Balkans markings, this machine is believed to have been photographed on Crete in 1941. The damage seems to have been the result of a particularly heavy landing which wrenched the starboard engine from its mounting and broke the aircraft's back. Note that although the fuselage carries the current 70/71 uppersurface colours, the starboard wing is still in the pre-war 61/62/63 scheme.

RIGHT: German paratroopers observing a crash-landed Ju 52/3m on Crete. Losses in men, gliders and aircraft during the Crete operation were very heavy and 271 Ju 52/3ms were destroyed or had to be written off as a consequence of landing on rough ground, on beaches, or in the face of intense gunfire. This was especially serious as the invasion of Russia was due to commence within one month.

LEFT: Once Crete had fallen, regular supply flights were still necessary to augment supplies arriving by sea. Here, Ju 52/3ms with yellow engine cowlings and tail surfaces are seen prior to taking more supplies to Crete in June 1941. Although the leaping stag emblem identifies these aircraft as having belonged to KGzbV 106, they may have been serving with either KGzbV 40, KGzbV 50 or KGzbV 60 when photographed.

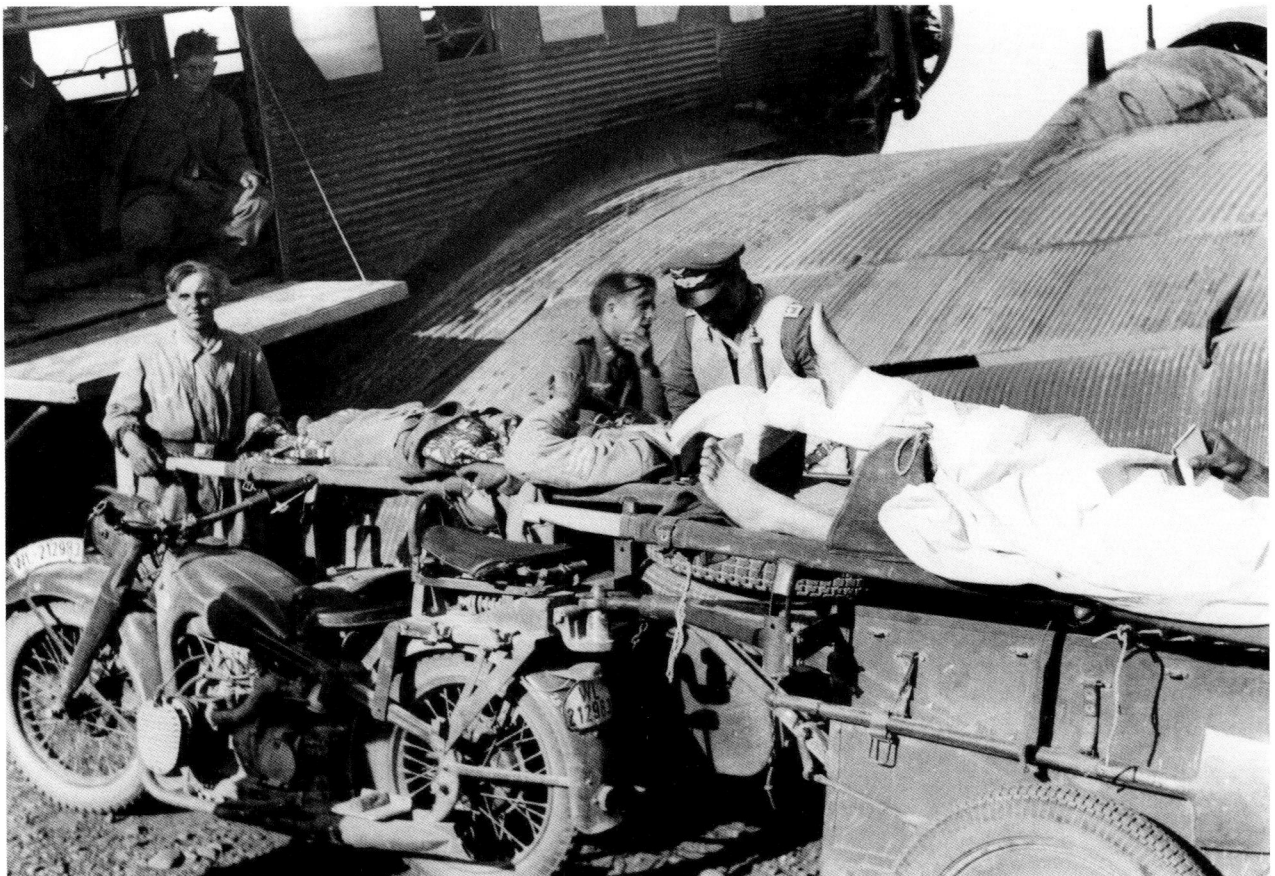

RIGHT: Another important task for the transport aircraft was the evacuation of wounded, seen proceeding here under what could almost be peacetime conditions. The motorcycle and sidecar, and the trailer upon which the stretchers have been placed, would all have been air lifted earlier to Crete.

1939-1943

RIGHT: The Staffel emblem on the nose of this Ju 52/3m is again the leaping stag of KGzbV 106 although, for the Crete operation, aircraft were withdrawn from this unit to form KGzbV 40, 50 and 60. The Staffel letter W on this particular machine suggests it was operating with KGzbV 60 when photographed.

BELOW: A Ju 52/3m in the Mediterranean theatre at the end of June 1941, at which time the aircraft was operating under the control of KGzbV 2. Note the yellow Balkans markings and the number 3 painted on the rudder.

The Eastern Front, June 1941 to April 1943

Even before Germany opened its attack on Russia on 22 June 1941, the transport difficulties were already recognised. With roads in poor condition and a rail system which had to be changed to the German gauge before it could effectively be utilised, the obvious solution was to call upon German air transport. The transport fleet, however, had suffered considerable losses in the Crete campaign which, moreover, had barely ended.

ABOVE: Ju 52/3ms in Russia, where the role of air transport was to be of the utmost importance. Note the tail flash on the machine in the foreground.

ABOVE: A decline in the strength and serviceability of the Ju 52/3m fleet soon after the beginning of the Russian campaign led to the creation of special glider units trained to supply armoured formations. This photograph shows Go 242s which have landed just behind the front and close to an armoured spearhead.

The *Lufttransportführer* was therefore faced with the impossible task of re-equipping the transport units recalled from the Mediterranean within the space of ten days. In the event, when the Russian campaign began, only IV./KGzbV 1, KGrzbV 50, KGrzbV 102 and KGrzbV 106 were available, while I./LLG 1, I. and II./KGzbV 1, KGrzbV 101, KGrzbV 104 and KGrzbV 105 were still refitting in Germany. Eventually, these *Gruppen* were allocated to the various *Luftflotten* as they advanced further into Russia, operating from bases located over the entire front.

By the end of September, the distances flown and the poor servicing facilities behind the front had so greatly reduced the strength and serviceability of the Ju 52/3m fleet that special glider units were created to meet the need for more transport aircraft. Equipped with the new Go 242 and known as 'Goliath' *Staffeln*, these units were specially trained to supply armoured formations. These gliders proved most satisfactory as they were simple to load and their large holds could accommodate up to 3,500 kg in food, fuel drums or medical supplies as well as ammunition of all calibres up to and including shells for heavy field howitzers. Even in bad weather and with a cloud base of some 183 metres it was possible for the gliders to land at a prescribed landing ground.

On 17 December, as the Russian winter offensive began to press hard on Army Group Centre, Hitler ordered the creation of five new transport *Gruppen*. These units, designated KGrzbV 700, KGrzbV 800, KGrzbV 900 and KGrzbV 999, were immediately despatched to VIII. *Fliegerkorps* in the central sector where all units under this command transported troops to reinforce the front.

Meanwhile, the Russians had broken through in the northern sector so that by 15 January 1942,

RIGHT: Some Ju 86 units were called up in February 1942 to participate in operations in the East but were found unsuitable and withdrawn early. This Ju 86, photographed earlier at Zeltweg-Steiermark in Austria, belonged to FFS (C) 11.

two *Gruppen* of VIII. *Fliegerkorps* in the central sector, KGrzbV 9 and KGrzbV 172, had to be sent to *Luftflotte* 1. They were joined in February by KGrzbV 500 and two further *Gruppen* raised from training schools designated KGrzbV 'Oels' and KGrzbV 'Posen'. The removal of five further *Gruppen* from the central sector in February depleted the transport resources of VIII. *Fliegerkorps* still further and by mid-February the situation had become so desperate that five more *Gruppen*, numbered KGrzbV 4 to KGrzbV 8, had to be formed for the transport of supplies to Army Group Centre and VIII. *Fliegerkorps*. However, KGrzbV 6 and KGrzbV 7 were equipped with Ju 86 and He 111 aircraft, types not particularly suitable for transport work, and they were sent back to the schools in March.

Apart from the normal supplies of fuel, lubricating oils, ammunition and tank and aircraft spare parts required by various *Wehrmacht* units on the Eastern Front, other loads varied from mines to headquarters staffs. In August, complete aircraft engines were taken to various destinations by Go 242s, and in October, pilots and essential ground personnel of I./JG 53 were transported from Stalino to Bari in Italy.

While VIII. *Fliegerkorps* was struggling to supply Army Group Centre, the breakthrough in the area of the front controlled by Army Group North had resulted in the encirclement at Demyansk of six divisions of the II. and X. *Armeekorps*. In order that these forces should hold out until relief was possible, it was proposed that all necessary supplies of weapons, ammunition, equipment, spare parts, food and replacement personnel would be flown in. The daily supply demand for the 100,000 men encircled at Demyansk amounted to 300 tons and to transport this at least 150 operational aircraft had to be available each day. However, as a result of their demanding missions in the central sector, the majority of aircraft assembled were in need of repair or overhaul. Indeed, many were already in the rear area repair shops so that only some 30 per cent of the aircraft belonging to the eight *Gruppen* due to participate were immediately available. It was therefore decided that more aircraft were needed to ensure that the daily 300 tons were delivered, and by early March another eight *Gruppen* were flying missions to Demyansk. The operation was the *Luftwaffe's* first large-scale air-supply undertaking of the war.

Units Participating in the Demyansk Operation

Unit	Commander	Previous Location	Base
KGrzbV 9	*Obstlt.* Johannes Jansen }		
	later *Obstlt.* Adolf Jäckel }	already in the area	Pskov South
I./KGzbV 172	*Major* Walter Hammer, later }		
	Hptm. Zähr }		
IV./KGzbV 1	*Hptm.* Fridolin Fath	Smolensk	Ostrov South
KGrzbV 500	*Major* Theodor Beckmann	Mediterranean	Pskov West
KGrzbV 600	*Hptm.* Zeidler (?)	Orsha	Korovye Selo
KGrzbV 700	*Major* Muggenthaler	Orsha	Pskov West
KGrzbV 800	*Major* Kalepke	Vitebsk	Korovye Selo
KGrzbV 900	*Hptm.* Stypschütz	Vitebsk	Pskov West
KGrzbV 999	unknown	Vitebsk	Pskov West
KGrzbV 'Posen'	under IV./KGzbV 1	Posen	Ostrov South
KGrzbV 'Oels'	*Obstlt.* Schweitzer	Oels	Pskov South
II./KGzbV 1 (part only)	*Hptm.* Riechers	Dnepropetrovsk	Ostrov South
II./KGzbV 1 (part only)	*Obstlt.* Neundlinger	Dnepropetrovsk	Ostrov South
KGrzbV 105 (part only)	*Obstlt.* Deffner	Vitebsk	Pskov South
KGrzbV 4	*Major* Rudolf Kraus }		Riga
KGrzbV 5	*Hptm.* Zahn, later }	from communications	
	Hptm. Fritz Uhl }	flights, schools, etc	Riga
KGrzbV 8	*Obstlt.* Damm }		Daugavpils

Note 1: Two further units, KGrzbV 6 and KGrzbV 7, were activated but soon dissolved. Their personnel were returned to training and their aircraft distributed among the remaining units.

Note 2: The officers shown in the table were the leaders of the respective units, or parts of units, actually taking part in the Demyansk airlift and were not necessarily the *Kommandeur* or *Kommodore*. This particularly applied when only parts of a unit were involved, and such detachments may have been under the command of an officer other than the *Kommandeur* or *Kommodore*. Thus, for example, the parts of KGrzbV 105 employed in the operation were under the command of *Obstlt.* Deffner, whereas the rest of the unit remained under the *Kommandeur*, *Major* Reinhold Wenning.

At first, the training and experience of the air transport personnel varied enormously, from older crews who had taken part in many previous operations, to newcomers who had only just completed their flight training. There was also a number of crews made up from instructors and personnel who had been wounded on earlier missions and who were returning to duty with units raised from various training establishments. However, once the younger crews had flown a few missions, they quickly acquired the necessary skills to become fully operational and the initial gulf between the experienced and inexperienced was narrowed.

Facilities and equipment at the bases used also varied, so that while Pskov South was a major base with full services, Pskov West was normally only an emergency airfield and lacked technical support and accommodation. Similarly, Riga was a commercial airfield adapted for military use, but Daugavpils was a minor airfield with limited services and Korovye Selo lacked hangars, runways and had only temporary barracks.

The landing base at Demyansk itself lacked any airfield facilities and consisted only of a landing strip with small taxiing and unloading areas. During the early stages of the operation, these were merely areas of levelled and hard-packed snow which had to be kept clear of fresh falls at all times. A strict timetable was laid down so that units could follow each other as quickly as possible to land, unload and take off for the return flight, often overloaded with wounded, without concentrating too many aircraft over the field at once. In March, a second strip was cleared south-east of Demyansk but it could cope with no more than six aircraft at a time and could be used by only the most experienced pilots.

Initial technical difficulties in temperatures often as low as minus 40 degrees C were overcome by setting up special aircraft servicing companies at key points under the command of experienced technical officers and engineers. The supply of spare parts was improved by extending the frequency of major overhauls from 450 to 500 flying hours, and by carrying out maintenance at the front instead of in the rear area it was possible to keep operational readiness at between 50 and 80 per cent. Once the operation was well under way, as many as 600 transport sorties were being flown daily.

As the operation continued into the thaw and muddy period, conditions deteriorated in the taxiing and landing areas at the take-off bases and to an even greater extent at the landing bases. Delays in loading, unloading and taxiing became inevitable and any flying schedule was quickly abandoned. Instead, aircraft took off in pairs or in groups of up to six machines whenever they were ready. For the return flight over the critical area near the front, all transports, regardless of their unit, assembled into groups and for their own protection stayed together until they had crossed the front line. They then separated and returned to their own particular take-off bases.

During the final phase of the operation, airfields dried out and conditions improved. By this time, servicing facilities at the take-off bases had also improved and the degree of operational readiness increased. It therefore became possible to reinstate and maintain a timetable and operations were aided by the longer hours of daylight. Flights were later flown in *Gruppe* formation at low altitude and were always led by the *Gruppenkommandeur* or, in exceptional circumstances, by a *Staffelkapitän*. Despite strong enemy opposition, losses were reduced by constantly changing the flight path and providing the transports with a small number of escort fighters from I./JG 51 and 9./JG 54. However, Soviet air power in the area was comparatively weak and the transports' concentrated firepower, especially if tracer ammunition was used, usually deterred attacks and some of the air-gunners aboard the transport aircraft succeeded in shooting down Soviet fighters.

Fortunately, the Soviets did not possess the means to attack the take-off bases. Had they been able to do so, they may well have destroyed significant numbers of aircraft either awaiting loading or lined up to take off and the entire operation may have been ruined for lack of replacement aircraft. As it was, the greatest danger was from ground fire; as soon as the Soviets learned of the supply sorties, every Soviet soldier was ordered to carry a weapon at all times and to fire at transport aircraft passing overhead. Crews reported being fired on with all manner of weapons, even flare pistols, and one Ju 52/3m crashed when its pilot, flying at low altitude, was seriously wounded by sub-machine gun fire. Particularly effective, however, were the Soviet self-propelled 20 mm anti-aircraft guns, since these could be moved to cover the transports' ever-changing routes, and the number of aircraft lost or badly damaged increased accordingly. There was, therefore, a steady increase in the number of aircraft that had to be taken out of action for repairs.

In some exceptional cases, the crews of aircraft which had been shot down and crash-landed in enemy territory were able to make their way back to safety after weeks of gruelling effort. More frequently, other pilots risked landing in open terrain to rescue their stranded comrades.

By overcoming difficulties arising from hasty planning, and with a tremendous expenditure of material and effort, the transport force was able to continue the airlift until a narrow land route to the Demyansk pocket was opened on 16 May. This connection, however, was unreliable, and Demyansk continued for the most part to depend on supply by air, so that by the 20th, when the majority of the formations could be transferred to other sectors of the front, or back to the Reich, three *Gruppen* were retained. From January 1942 to the final clearing of the Demyansk area in early 1943, the transports flew a total of 33,086 sorties during which 64,844 tons of materials, supplies and equipment and 30,500 troops were flown in, while on their return trips, the transports evacuated 35,400, most of whom were wounded or sick.

Meanwhile, the transport forces had also been supplying the much smaller force of 3,500 German troops trapped at Kholm. Unlike Demyansk, a factor of this operation was that the enemy gradually moved ever closer to the airfield. The Soviets realised that if the landing ground was captured, the ground force would be cut off from its only source of supply and, despite stubborn German resistance, slowly gained ground. This made the mission of the air supply units progressively more difficult as the only suitable landing area soon came within range of Soviet artillery. On one occasion, seven Ju 52/3ms forced a landing under fire, but five were immediately destroyed and no further landings were attempted. Instead, supply was conducted using gliders: the He 111s of KG 4 towing Go 242s while the Ju 52/3ms of II./KGzbV 1 towed DFS 230s. Twilight operations were carried out at dawn and dusk in order to minimise losses to artillery fire and, after releasing their Go 242s, the He 111s tried their best to hold down enemy fire while the gliders landed in a street in Kholm. Even so, losses were high and, as the gliders could not be retrieved for reuse, such operations were suspended in favour of supply drops. When the Kholm action ended on 20 May, a total of 24,303 tons of supplies and 15,445 troops had been carried into the pocket.

Although successful, the Demyansk and Kholm operations had involved the commitment of almost the entire air transport force and the Demyansk action was completed at the cost of 265 Ju 52/3ms lost due to enemy action, crashes or emergency landings, and 383 personnel killed, wounded or missing. Moreover, while the training schools were deprived of 300 aircraft for four months, the success of the airlifts set a dangerous precedent for future *Luftwaffe* planning, and undoubtedly influenced the decision to supply the Sixth Army at Stalingrad.

RIGHT: Ground personnel on a forward airfield in Russia in the summer of 1941 preparing to load 200 litre barrels of petrol or oil into a Ju 52/3m. This machine, coded B1+EA, was flown by Transport Staffel/I. Flieger-korps and had black or dark green undersurfaces. The striped cowling was typical of the Staffel's aircraft but the significance of the letter H on the forward fuselage is not known.

LEFT: Horse-drawn transport being used to carry recently delivered supplies to their onward destination. The Ju 52/3m was W.Nr. 7060 and had the Stammkennzeichen KB+RS in black, narrowly outlined in white.

BELOW: Unloading supplies from a Ju 52/3m. Although the Transport Gruppen did their utmost to meet the requests of the Army for airborne supplies, this could not always be carried out in full due to bad weather or because the number of aircraft available was too small in relation to the size of the units to be supplied. In extreme weather, loads often contained fodder to sustain the horses widely used by the Wehrmacht.

ABOVE AND BELOW: The temporary, composite Gruppen KGrzbV 40, KGrzbV 50 and KGrzbV 60 were disbanded immediately after 'Merkur' and their aircraft transferred as replacements to units which had experienced greater losses. Thus, although the unit code 9P+EL on this machine, seen in Finland in September 1941, indicates that it had earlier served with KGrzbV 40, the leaping stag emblem on the nose indicates that, when photographed, it had been assigned to KGrzbV 106.

A similar aircraft, also of
KGrzbV 106 in Finland in
September 1941, was coded
9P+DD and had evidently
come from the disbanded
KGrzbV 60.

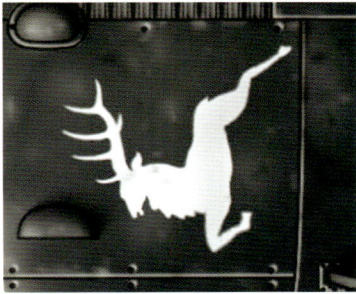

Emblem of KGrzbV 106

Ju 52/3m coded 9P+DD of Stab/KGrzbV 106, Finland, September 1941
This machine was finished in a standard yet rather weathered 70/71/65 scheme but had recently been fitted with a replacement rudder, tailplanes and engines. The door was also new or recently repainted. Although the lower part of a yellow theatre band was applied to the fuselage, the yellow under the new engine cowlings had yet to be applied. The full operational code was applied under the wings with the individual aircraft letter D repeated on the leading edges, and an indecipherable Werknummer appeared on the fin in white.

1939-1943

RIGHT: Assisting operational units move location quickly and efficiently was particularly important during the early months of the Russian campaign when frequent transfers were required in order to keep pace with the speed of the advance. Here, a Ju 52/3m is accompanied on a transfer flight by Ju 87 Bs of St.G 2.

LEFT: A Ju 52/3m in Russia showing the emblem of I./KGzbV 1. More specifically, the operational code IZ+NK on the fuselage identifies the machine as belonging to 1. Staffel.

RIGHT: This orderly scene shows wounded German soldiers being offloaded from Ju 52/3m transports early in the Russian campaign. Such conditions could not be maintained, however, and deteriorated as the war progressed. The palliasses were already standard when this photograph was taken, probably in the summer of 1942, but within a year, paper dressings and blankets had replaced those seen here.

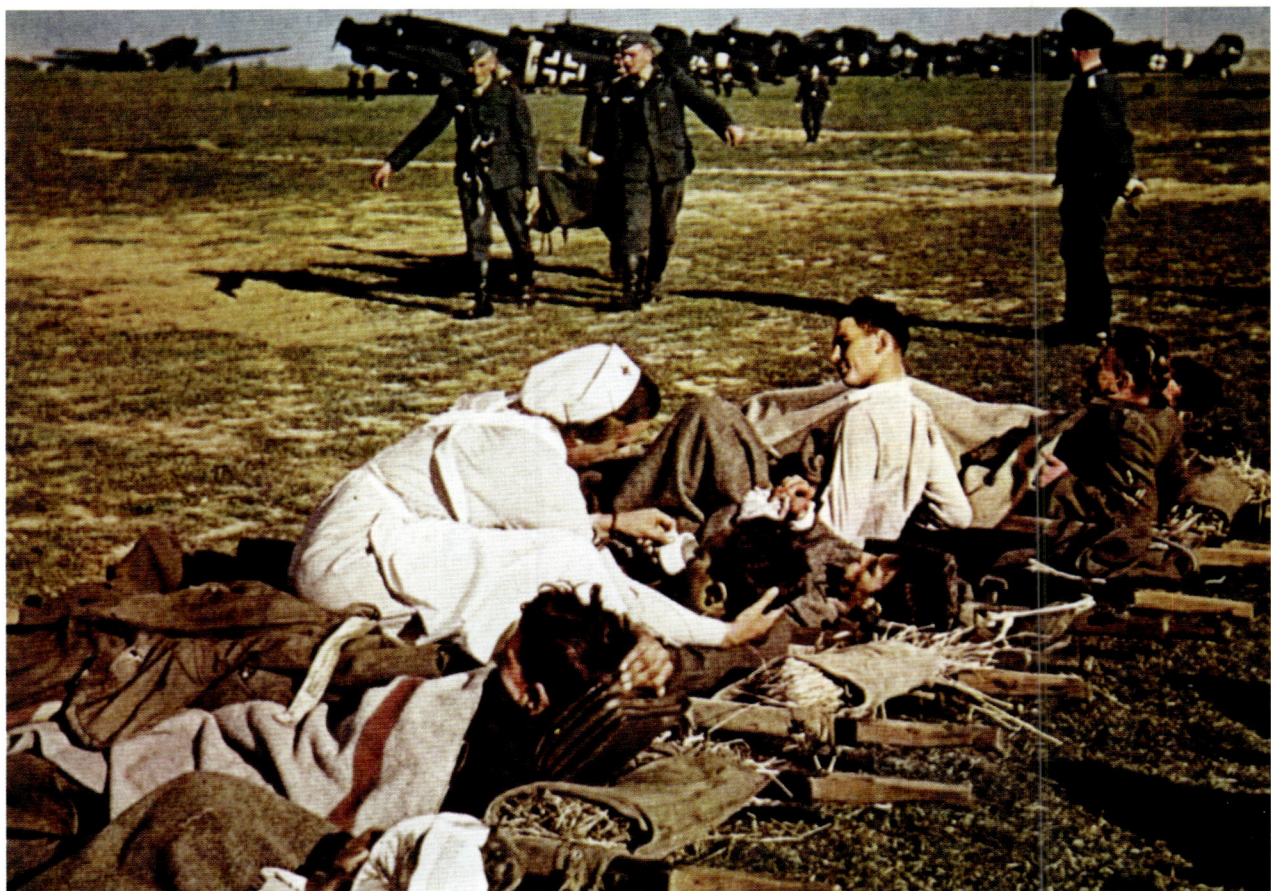

LEFT: Ju 52/3ms of 1./KGrzbV 1 at Zaporozhye in southern Russia in the autumn of 1941. Points of interest include the severe weathering and faded paintwork on the aircraft nearest the camera, and the exhaust staining and individual aircraft letter K on the wing of the machine next in line.

BELOW: A Ju 52/3m raises the dust as the pilot increases the power of the engines before commencing his take-off roll. The yellow lightning bolt on the forward fuselage suggests the machine may have belonged to an air signals detachment.

LEFT: With the onset of winter, braking and manoeuvring in the ice and snow was difficult and a good landing was considered quite a feat. This Ju 52/3m, 1Z+MK of 2./KGzbV 1, ended up on its nose after a landing accident at Zaporozhye in late 1941 when the pilot braked too hard. Interestingly, although the numeral 1 appears on the yellow undersurface of the wing tip, indicating that the complete code was painted under the wings, there is no sign of the letter Z, despite the fact that the reflected light from the snow has illuminated the white of the underwing Balkenkreuz. This would suggest that the undersurfaces have been painted later, almost certainly in black. The yellow fuselage band on this aircraft has been applied further forward than was usual.

ABOVE: Probably photographed in the winter of 1941/42, this Ju 52/3m belonged to Stab, I./KGzbV 172, and took part in the Demyansk operation. Note the extensive weathering and the aircraft letter D repeated under the wing.

BELOW: Almost certainly taken in the Demyansk pocket, this photograph shows supplies being unloaded from a winter camouflaged Ju 52/3m. The machine carries a Stammkennzeichen on its fuselage sides and, having been called up from a training unit, is believed to have then been allocated to KGzbV 700.

One of the Ju 52/3m aircraft which took part in the Demyansk airlift was this machine, coded 4V+CM, in a weathered snow camouflage and with a glider-towing attachment on the tailwheel. This aircraft, W.Nr. 7515, was shot down by Soviet anti-aircraft fire near Stalingrad on 26 December 1942 and all four members of the crew were killed.

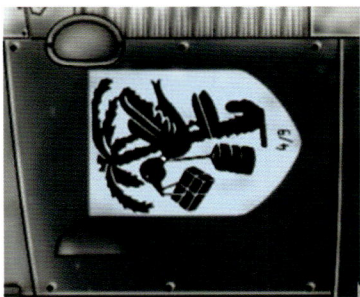

Emblem of 4./KGrzbV 9.

Junkers Ju 52/3m of 4./KGrzbV 9, Russia, early 1942
With the exception of some partial overpainting on the upper part of the yellow fuselage band, the temporary white winter scheme on this machine had originally been carefully applied, particularly around the fuselage lettering. However, subsequent weathering has eroded the white and exposed areas of the original 70/71 green finish, particularly on the leading edges of the wings and tail surfaces. Note the handrail mounted under the fuselage windows and that the area of the nose showing the Staffel emblem was not winter camouflaged.

1939-1943

ABOVE AND BELOW: These aircraft belonged to KGrzbV 9 and were photographed during the Demyansk airlift. Furthest from the camera, (*ABOVE*) is a snow camouflaged machine from a training establishment showing the Stammkennzeichen VB+UF, while the closer machine is W.Nr. 6682 coded 4V+AK. This latter aircraft belonged to 2./KGrzbV 9, had the Staffel emblem on the nose and the individual aircraft letter A narrowly outlined in red. Note the yellow wingtip and the letter A repeated outboard of the wing Balkenkreuz.

ABOVE: A low-flying Ju 52/3m of KGrzbV 9 taking supplies into the Demyansk pocket and showing the unit number on its rudder. During the entire operation, aircraft consumed 160 train-loads of fuel, 265 Ju 52/3ms were destroyed and the pilot training programme was deprived of 300 aircraft for a period of four months. In fact, the effects on training were so catastrophic that they had still not been overcome at the end of the war. Demyansk proved also to be a rather dangerous example of the use of air transport, its success encouraging the belief that such operations could be employed in any similar situation in the future.

ABOVE: The main landing base in the Demyansk pocket consisted only of a landing strip with small taxiing and unloading areas. During the early stages of the operation, these consisted of hard-packed snow, but in the thaw and muddy period, conditions deteriorated badly. As shown here, however, in the later stages of the operation in April and May 1942, conditions improved once the airfield had dried out. Some of these Ju 52/3ms belonged to I./KGzbV 172, the rudder of the aircraft on the far left showing the Gruppe's star and lucky pig emblem, below which is the Staffel number.

LEFT: The only two members of KGzbV 172 to receive the Ritterkreuz were Fw. Oskar Kräussel, (*FAR LEFT*) later killed in March 1943, and Major Erich Zähr, who was Kommandeur of I. Gruppe in 1942 and 1943. Both received the award on 24 December 1942.

1939-1943

RIGHT: A winter camouflaged Ju 52/3m at Zaporozhye in southern Russia in the winter of 1941/42. This particular machine, 1Z+AL, was flown by Uffz. Friedrich Hesse of 3./KGzbV 1.

BELOW: This Ju 52/3m of an unidentified unit in Russia appears to be finished overall in green 70 or 71 with IP, the last two letters of the Stammkennzeichen seen under the wing, repeated on the leading edge.

LEFT: This aircraft was finished in a temporary white winter camouflage that had weathered to the extent that it has been completely eroded from the nose area. It would seem from the emblem on the nose that this aircraft had once belonged to Stab/KGzbV 105, but the new operational code G6+EZ shows that the machine had later been allocated to 3./KGrzbV 105. The E in the code and the band around the rear fuselage were yellow, and the areas of fresh paint either side of the fuselage Balkenkreuz show where the earlier Stab markings have been painted out and replaced with those of 3. Staffel. Another area of fresh paint on the rudder shows where a tactical code has been deleted. Parts of KGzbV 105 were employed during the Demyansk operation.

BELOW: Comparatively few Ju 52/3ms were fitted with skis, as seen on this machine which carries the code P4+FH of Transport Staffel/Fliegerführer Nord (Ost), probably in early 1942.

1939-1943

LEFT: The arms of the city of Dortmund on the nose of this Ju 52/3m in Russia in early 1942 identifies the aircraft as belonging to KGrzbV 500. This Gruppe had been transferred from the Mediterranean in order to participate in the Demyansk airlift.

BELOW: This Ju 52/3m coded IZ+AK was assigned to the Staffelkapitän of 2./KGzbV 1 and is shown near Smolensk in 1942.

ABOVE: A Ju 52/3m undergoing maintenance in a well-appointed hangar in Norway. Paradoxically, units in the far north were better equipped to withstand harsh winter conditions than units operating in southern Russia where men and machines suffered greatly from the effects of the extreme cold.

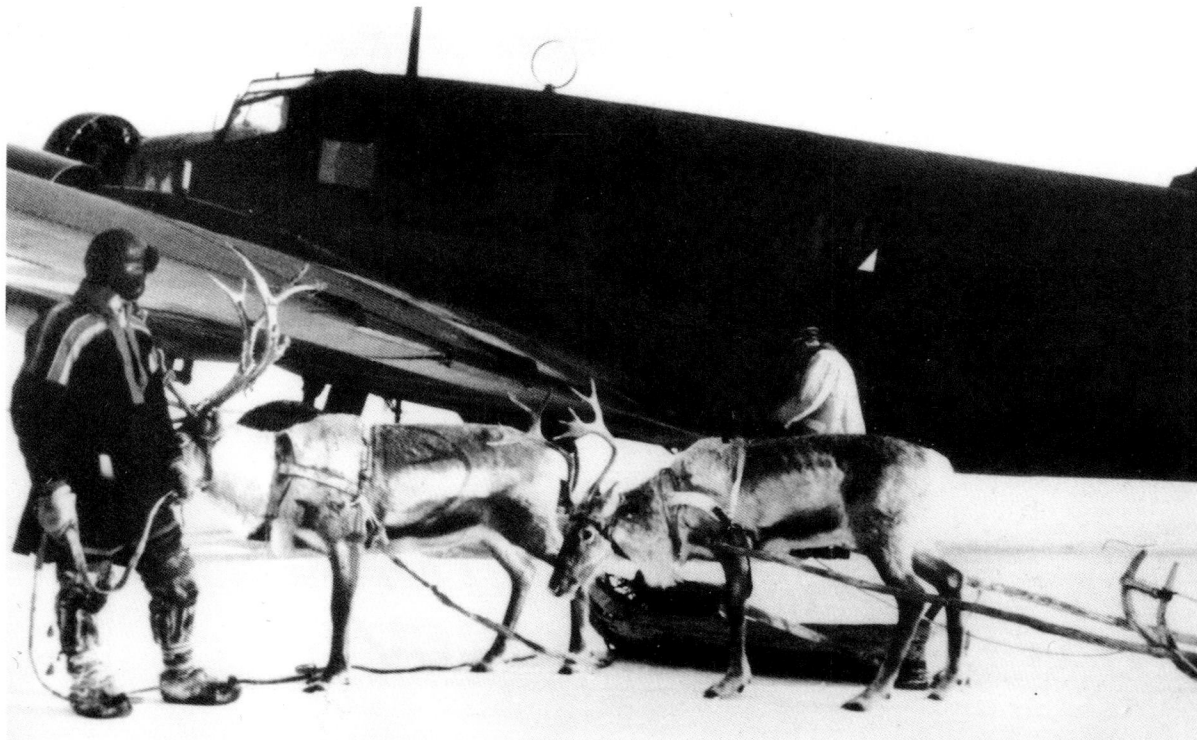

LEFT: A Ju 52/3m of KGrzbV 108 at Kirkenes in the far north of Norway, close to the borders between Russia and Finland, in late 1942 or early 1943. In May 1943, KGrzbV 108 was redesignated TGr 20 and maintained regular flights to Kirkenes until late in the war.

1939-1943

RIGHT: Showing signs of wear and tear, this winter camouflaged Ju 52/3m has the later type of double-shield emblem adopted by II./KGzbV 1 in mid-1941. The number 6119 just forward of the wing root is almost certainly the aircraft's Werknummer, and although in an unusual location, similar numbers have been observed on various other Ju 52/3ms.

BELOW: The cargo door on the side of the Ju 52/3m was large enough to permit the loading of bulky items. Taken at Bagerowo in the Crimea in early 1943, this view shows a field kitchen being manhandled into position.

LEFT AND ABOVE: With its centre and port engines already running, thick smoke belches from the exhausts of this Ju 52/3m as the starboard engine is fired up. The two views show what is believed to be the same Ju 52/3m, 4V+GK, of 2./KGrzbV 9. The individual aircraft letter G was painted on the leading edge of each wing and the Staffel emblem was applied to both sides of the forward fuselage. After being formed in August 1939, KGrzbV 9 had a distinguished history and served on the Western Front, in Russia and in the Mediterranean. Three of its aircrew were awarded the Ritterkreuz and at least 25 received the German Cross in Gold. The Gruppe was later redesignated I./TG 3.

ABOVE, ABOVE RIGHT AND RIGHT: The emblems of some of the transport units are shown here and include the running hare of 15./KGzbV 1 (*ABOVE*), the devil riding a bomb, as used by IV./KGzbV 1 (*ABOVE RIGHT*), and a blue shield featuring a bird carrying a packing crate and an oil drum, as used by 4./KGrzbV 9 (*RIGHT*).

LEFT: Just visible on the nose of this Ju 52/3m is the lucky pig emblem of 3./KGzbV 172.

RIGHT: A close-up view of the emblem of 3./KGrzbV 105 showing the Staffel number in yellow between the wings. The Stab emblem was similar but had the word 'Stab' between the wings and included a banner across the lower front of the barrel that read 'Narvik-Kreta', the unit's battle honours.

The Air Supply of Sixth Army at Stalingrad

On 11 June 1942, the OKL's *Lufttransportführer Ost*, *Oberst* Friedrich-Wilhelm Morzik, was ordered to set up a headquarters with *Luftflotte* 4 to supply Sixth Army and VIII. *Fliegerkorps* advancing in the summer offensive against Stalingrad and Fourth *Panzer Armee's* advance into the Caucasus. For this purpose, the four Ju 52/3m *Gruppen* and one He 111 *Gruppe* already with *Luftflotte* 4 were reinforced by five more Ju 52/3m *Gruppen*, these units being based on airfields east of the Dnieper Bend and north of the Sea of Azov. All manner of Army and *Luftwaffe* supplies were transported to the front including ammunition, aviation and vehicle fuel, aircraft parts and Army freight, while more than 51,000 wounded were evacuated. Thus, by mid-November, the requirement for air transport and the scale of effort was already abnormally high, with the result that operational readiness had fallen to 40 per cent and a number of *Gruppen* had to be withdrawn to Germany to refit. Similarly, the German troops, committed without respite since the beginning of the offensive, were exhausted, and their tanks, vehicles, weapons and all types of equipment were worn out. The rapid advance had not allowed a proper transport channel to be established and the air transport force was already deeply committed to making up this deficiency when Sixth Army advanced into the outskirts of Stalingrad itself.

Then, on 19 and 20 November, and coinciding with the first signs of winter, the Soviets boldly counter-attacked to the north-west and south of Stalingrad, broke through the front and threatened to encircle Sixth Army. Hitler was informed of this critical turn of events at the Berghof in Berchtesgaden on the 19th, and despite continual updates throughout that afternoon, believed that the Soviet breakthrough could be contained and that the threat to Sixth Army would exist only for a few days. On the 20th, Hitler sent for the *Luftwaffe* Chief of Staff, *General* Hans Jeschonnek, to explore the role the *Luftwaffe* might play in any relief operation.

Once Hitler had explained that the encirclement of Sixth Army was temporary and that he believed it was possible to break the encirclement and restore the position on the southern front, Jeschonnek assured him that if bomber as well as transport aircraft were employed, and if suitable airfields could be maintained, the *Luftwaffe* could airlift sufficient supplies. It will be noted that, at this stage, *Reichsmarschall* Hermann Göring had not been consulted. He was, in fact, presiding over an oil conference at Karinhall near Berlin but, in any event, Hitler, having realised that Göring was not keeping him properly informed, now preferred to discuss *Luftwaffe* matters with Jeschonnek.

The next day, Hitler ordered Sixth Army to hold its positions, adding that special orders regarding air supply would follow. Only then was Göring informed of the decision, but instead of demanding statistical data to support the feasibility of the airlift, Göring, like almost everyone else directly involved at higher command level, assumed that the encirclement would be temporary and, supporting Jeschonnek, merely assured Hitler that the *Luftwaffe* would do all it could to meet the Army's needs.

On the 22nd, the encirclement of Sixth Army was complete and its 300,000 men were practically cut off from all sources of supply. The Army's supply situation, already critical, was bound to become catastrophic within a few days. Hitler, however, had already made up his mind that for political and military reasons, Sixth Army would remain at Stalingrad and that the city was to be designated a 'Fortress.' Thus, when he met Göring at the Berghof, Hitler had before him figures prepared by Jeschonnek and the *Reichsmarschall* was informed that, if the *Luftwaffe* could not carry out the task of air-supply, the Army was lost. As Göring later remarked, when presented in this fashion, Hitler had him "firmly by the sword-knot" and he could therefore only agree. Thus the fate of Sixth Army was sealed.

Jeschonnek, however, soon discovered that his original calculations were incorrect, and although he informed Göring of his error, he was ordered by Göring not to inform Hitler. Nor did Göring himself enlighten Hitler. On the contrary, he left for Paris to purchase works of art.

The Army itself, having been ordered to hold fast until a relief operation could be mounted, calculated that it would require 750 tons of supplies per day to survive. Although the earlier Demyansk operation had already illustrated the potential usefulness of air transport, neither Hitler, Jeschonnek nor Göring yet envisaged an airlift on anything like the same scale or duration as at Demyansk. The Army's opinion, however, was that the trapped forces would have to be supplied by air for weeks, possibly for months, although its estimate of 750 tons per day was revised to 500 tons. Later

Friedrich-Wilhelm Morzik was born in 10 December 1891 at Passenheim in East Prussia. Morzik flew as a Leutnant during the First World War and after joining the Luftwaffe, became Kommodore of KGzbV 1, which he led from 26 August 1939 to 1 August 1941. This photograph shows Oberst Morzik wearing the Ritterkreuz he was awarded on 16 April 1942 for his leadership of KGzbV 1 and his role as Lufttransportführer Ost with Luftflotte 1 during the Demyansk airlift.

ABOVE: At the time of the Stalingrad airlift, Luftflotte 4 was commanded by General Wolfram Freiherr von Richthofen, on the right, who held this position from July 1942 until September 1943.

calculations, first at Hitler's headquarters and then by the Army itself, showed that an absolute minimum of 300 tons per day would be required.

From the very beginning, *Luftflotte* 4 officials, to whom the mission had been assigned and who had been urging the trapped Army to break out, voiced their doubts and reservations. *General* Wolfram von Richthofen, commanding *Luftflotte 4*, was well aware that the summer offensive had already depleted the transport units and warned that as he then possessed only 80 transport aircraft, there was not a hope of supplying the Army, especially in the weather at Stalingrad. He declared the proposal *"Stark, staring madness!"* G*eneralleutnant* Martin Fiebig, commanding VIII. *Fliegerkorps*, was equally thunderstruck, exclaiming, *"A whole Army? It's quite impossible!"*

Everything did indeed depend upon the weather, but southern Russia was notoriously unpredictable in this respect and frustrated countless forecasts. Clear skies suddenly gave way to low cloud, while fog, sleet and snow alternated with only brief clear periods. Nevertheless, regular airlift operations began on 24 November and on the 28th, Richthofen appointed *Generalleutnant* Fiebig and his VIII. *Fliegerkorps* solely responsible for the air supply of Sixth Army. Immediately, measures were taken to reinforce *Luftflotte* 4's existing air transport strength by requisitioning all available Ju 52/3ms. KGrzbV 500 and KGrzbV 700 were brought up to full strength and by 1 December a programme had been drawn up for assembling at Morozovskaya and Tatsinskaya, east of the Lower Donets, the serviceable aircraft of ten Ju 52/3m *Gruppen*, one *Gruppe* of He 177s, and one mixed *Gruppe* of Fw 200s, Ju 90s and Ju 290s. Most of the Ju 52/3m *Gruppen* had been formed by concentrating all available aircraft from schools and workshops and flying them to *Luftflotte* 4. In addition, training and schools aircraft were formed into KGrzbV 20, 21 and 22 with He 111s and Ju 86s. Thus, within a few days, some 500 transport aircraft were ready and serviceable, and more aircraft would be made available as the operation progressed. Faults in ground and operational organisation, however, led to such a drop in serviceability that the transport units were never able to cope properly with the operation. A daily average of no less than 300 tons of supplies was needed to prevent Sixth Army from starving to death on the frozen steppe, but then the weather closed down. *'Weather atrocious'*, Fiebig wrote. *'We are trying to fly but it's impossible. One snowstorm succeeds another'*.

ABOVE: On 28 November 1942, von Richthofen appointed Generalleutnant Martin Fiebig (right), the commander of VIII. Fliegerkorps, with responsibility for the air supply of Sixth Army at Stalingrad.

Even when the weather cleared, overcrowding of bases, shortages of petrol and severe icing conditions created immense difficulties, with the result that in the five days from 26 to 30 November, a daily average of only 75 tons of supplies was flown into the Stalingrad pocket [5]. In the following nine days, this was increased to an average of 117 tons per day, but that was still little more than a third of the quantity that Sixth Army considered its minimum.

By 20 December 1942, the daily average of supplies reaching Stalingrad since the operation began had increased to 121 tons, but the trapped forces were still consuming more than they were receiving. In addition, the position regarding aircraft availability had become so desperate that three Ju 52/3m *Gruppen* had to be withdrawn from the Mediterranean, re-equipped at Berlin-Staaken with aircraft taken from *Lufthansa*, communications units, duty flights and even industrial and technical *Staffeln*, and sent to *Luftflotte* 4.

Some operational bomber units were also called upon to work alongside the *Transport Gruppen* and shared the task of carrying in supplies and evacuating the wounded. Apart from some He 111 *Gruppen* of KG 4, KG 27, KG 55 and KG 100, these included 20 Fw 200 long-range bombers from KG 40 which were brought to Zaporozhye where they joined the He 177s of the newly-formed I./KG 50. A full list of the units engaged in supplying Sixth Army appears in the following table:

---

5. The averages mentioned here have been calculated from the war diary of Sonderstab Milch. Von Richthofen's war diary gives contradictory figures, as do almost all published accounts of the Stalingrad airlift.

An He 177 photographed during the winter of 1942/43. Although this aircraft carried the badge of KG 40 on the nose, it almost certainly belonged to I./KG 50 which was then carrying out winter trials on the Eastern Front. When transferred to Zaporozhye to take part in the Stalingrad operation, the He 177s operated for the duration of the airlift under the designation Fernkampf-geschwader 2.

Units Employed in the Stalingrad Airlift

Units based at Tatsinskaya under
Oberst Hans Förster, Kommodore of KGzbV 1

KGrzbV 9	Ju 52/3m
KGrzbV 50	Ju 52/3m
KGrzbV 102	Ju 52/3m
KGrzbV 105	Ju 52/3m
I./KGzbV 172	Ju 52/3m
KGrzbV 500	Ju 52/3m
KGrzbV 700	Ju 52/3m
KGrzbV 900	Ju 52/3m
KGrzbV 21	Ju 86
KGrzbV 22	Ju 86

Units transferred from Italy, late December 1942

I./KGzbV 1	Ju 52/3m
II./KGzbV 1	Ju 52/3m

Units based at Morozovskaya under
Oberst Dr. Ernst Kühl, Kommodore of KG 55

KG 27 (two Gruppen)	He 111
II./KG 55	He 111
III./KG 55	He 111
KGzbV 5	He 111
KGzbV 20	He 111
KGzbV 23	He 111
III./KG 4	He 111
I./KG 100	He 111

Heavy Transport Units at Stalino and Zaporozhye

1. and 3./KG 40 [6]	Fw 200
One Staffel	Ju 90 and Ju 290
I./KG 50	He 177 [7]

When the weather permitted, the supply units flew in Staffel strength or in groups of five aircraft with fighter escort, but during bad weather only crews experienced in instrument flight flew in groups while the rest flew singly. Not surprisingly, crews straight from training schools were the least capable of dealing with the difficulties. Given no time to become gradually accustomed to the appalling conditions, they had to be fully employed from the start. The harshness of the winter cold; the long and dangerous approach and return flights over enemy territory bristling with anti-aircraft guns and fighter opposition; the constant enemy bombardment of take-off fields; the unloading and loading in the encircled area while constantly harassed by artillery fire and grenades; and the ever-present danger of icing and other technical failures in the bitter cold, all combined to make the situation exceedingly difficult for even the most experienced crews.

---

6. These Staffeln were amalgamated throughout the airlift and, together with the Staffel of Ju 90 and Ju 290 aircraft, operated as KGrzbV 200.
7. This Gruppe, which was carrying out winter trials on the Eastern Front when it was transferred to Zaporozhye to take part in the airlift, was redesignated Fernkampfgeschwader 2 (Long-Range Bomber Geschwader 2, abbreviated as FKGr 2) for the duration of the operation.

ABOVE: This aircraft was engaged in the supply of Stalingrad in late December 1942, when it was attacked by three Soviet fighters. The co-pilot was wounded in the attack and the machine was damaged to the extent that its starboard undercarriage collapsed on landing. It is believed the machine belonged to KGzbV 102.

Aircrews also had to endure the sight of their half-starved Army comrades in the encirclement area, some in uniforms so dirty, worn and disintegrating that they were barely recognisable as soldiers. After all the horses had been eaten, one stale loaf per day was the only ration available for every five men. All heating supplies had been exhausted, and as only one-third of the troops had bunkers, many slept in shell holes or on the frozen ground with only a single blanket to cover them. Consequently, there were countless cases of frostbite and thousands froze to death. Particularly distressing, however, was the plight of the vast numbers of wounded. Many were left in the open in long files to await evacuation, but without proper protection, they frequently succumbed to the cold. Even experienced crews often returned from these harrowing missions badly shaken by what they had witnessed. A crew from Transport Staffel/II.Fliegerkorps stated that they had seen a lorry packed with frozen corpses. A Feldwebel pilot of 3./KGzbV 1 discovered when evacuating wounded that their bandages and clothing were infested with lice and he found the fuselage of his machine stained with blood. Hptm. Werner Müller, the Staffelkapitän of 2./KGzbV 172, had seen soldiers piled on top of one another in houses to keep themselves from freezing to death and described the situation as the worst he had ever experienced. Some sick and wounded had to be left behind as there was insufficient space in the aircraft to fly them out, and in order to deter them from rushing the aircraft, loading had to be supervised by an officer with a pistol in his hand. Nevertheless, at the beginning of every flight, desperate, half-frozen men still tried to cling to the departing aircraft.

On 23 December, Russian tanks were observed moving up into position near the take-off base at Tatsinskaya. However, as Hitler had personally ordered that the base should not be evacuated until actually under fire, Luftflotte 4 was reluctant to abandon the field and would not grant General Fiebig permission to leave. Only when Russian tanks were actually roaming about the airfield on the 24th did Fiebig order his units out. With visibility down to less than 600 metres, the pilots took off amid indescribable confusion and under a steady hail of Russian fire. Aircraft scrambled from all corners of the airfield at once, one piloted by a Hauptmann from a Signals Regiment who had never flown an aircraft before in his life. Several machines crashed or collided in the chaos, with the result that while at least 108 of the original number

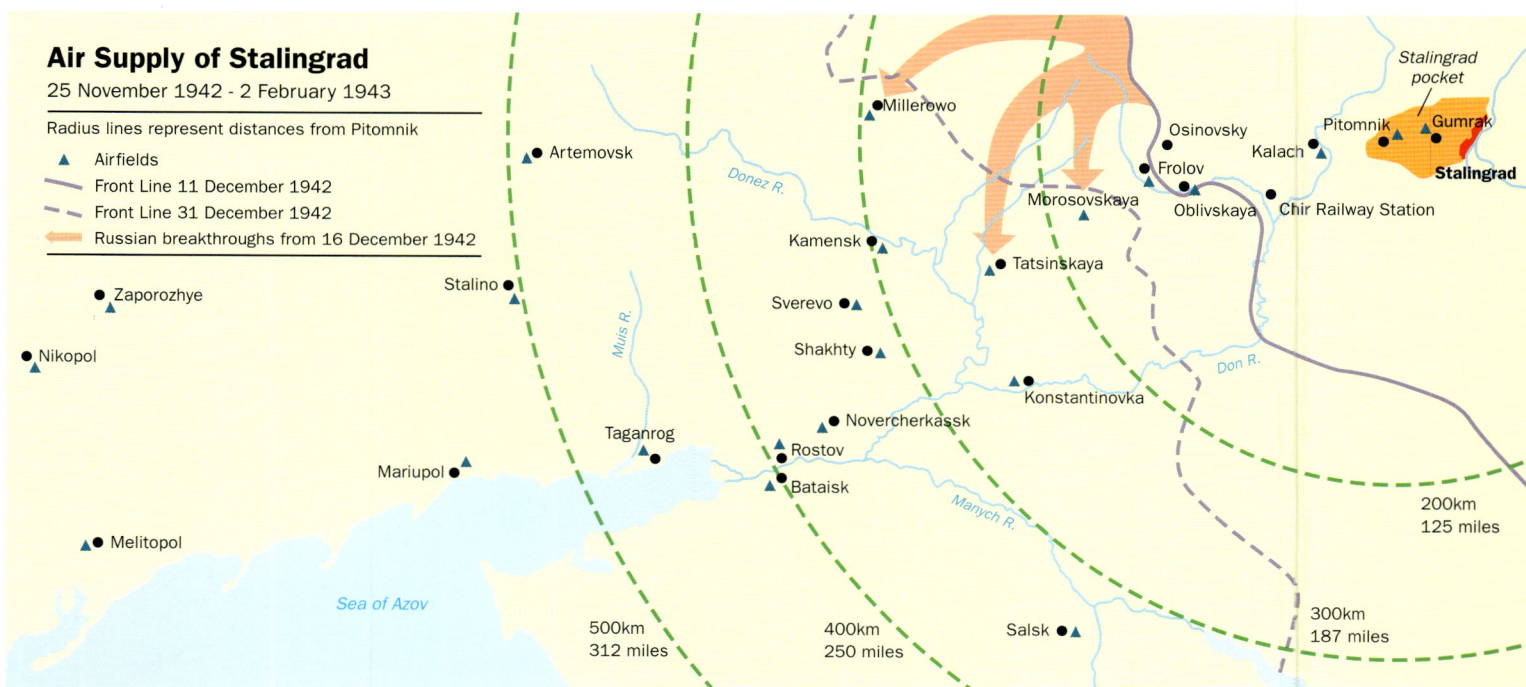

Air Supply of Stalingrad

25 November 1942 - 2 February 1943

Radius lines represent distances from Pitomnik

▲ Airfields
— Front Line 11 December 1942
-- Front Line 31 December 1942
➡ Russian breakthroughs from 16 December 1942

Millerowo
Osinovsky
Pitomnik
Gumrak
Kalach
Frolov
Stalingrad pocket
Stalingrad
Artemovsk
Donez R.
Morosovskaya
Oblivskaya
Chir Railway Station
Kamensk
Tatsinskaya
Zaporozhye
Stalino
Sverevo
Don R.
Nikopol
Shakhty
Muis R.
Konstantinovka
Taganrog
Novercherkassk
Rostov
Mariupol
Bataisk
200km 125 miles
Manych R.
Melitopol
Sea of Azov
500km 312 miles
400km 250 miles
Salsk
300km 187 miles

of Ju 52/3ms escaped, 46 aircraft were lost. Moreover, as most of the units' technical equipment had been abandoned in the hasty evacuation of Tatsinskaya, serviceability dropped alarmingly and in a single blow the operational readiness of the transport units was reduced to less than 25 per cent. Over the next seven days, the average daily tonnage of supplies delivered to Stalingrad, which had increased to 215 tons in the five days prior to the loss of Tatsinskaya, dropped to a pitiful 70 tons.

The Ju 52/3m units were then ordered to continue their missions from Salsk, in the northern Caucasus, which involved a flight of some 400 km. Heavy frosts, icy 80 km/h winds and snowstorms reduced operational readiness still further. Particularly hard hit were the units which had flown in from the Mediterranean. They had either long forgotten or had never been instructed in the recommended cold start procedure in which petrol was added to the engine oil. This altered the viscosity of the oil so that it flowed more freely and made aircraft engines easier to start. As the engines became hot, so the petrol evaporated and the proper oil viscosity was restored.

On 2 January 1943, the Soviet advances forced the He 111 units at Morozovskaya to move to Novercherkassk where they suffered the same problems as the Ju 52/3ms operating from Salsk. The flight from Novercherkassk to Pitomnik, the only properly equipped airfield within the beleaguered city of Stalingrad, was over 330 km, while Salsk was now almost at the limit of the Ju 52/3ms' range. In order to fly the increased distances, extra fuel had to be carried at the expense of cargo, and when aircraft consuming abnormal amounts of oil as a result of excessive piston wear had to be withdrawn from operations, the delivery of supplies decreased still further.

ABOVE: All He 111 units assembled for the air supply of Sixth Army. Operating in the transport role, they were placed under the command of Oberst Dr. Ernst Kühl, the Kommodore of KG 55, seen here as a Major inspecting Flak damage to the spinner of an He 111.

Soon the advanced airfield at Salsk was in danger of being overrun. Soviet artillery and strafing attacks destroyed ten Ju 52/3ms on 18 January, while a further 20 had to be sent off for repairs. Consequently, the Ju 52/3ms were moved to Zverevo where, without adequate billeting or servicing facilities, they operated from a cornfield barely within operational range. Within 24 hours of their arrival, 52 aircraft had been lost to a Russian bombing attack, while a number of heavy snowstorms restricted operations for days at a time and each surviving Ju 52/3m had to be laboriously dug out of the snow drifts before supply sorties could be resumed.

Within the encircled area itself, the troops' situation grew more critical from day to day. Army *Hauptmann* Winrich Behr wrote: *'No miracle on the steppe can help us here, only good old Aunt Ju and the He 111s if they come – and come often.'* On 14 January, in an attempt to improve the situation, Hitler ordered *Generalfeldmarschall* Erhard Milch to take over the airlift, but even he was unable to affect the outcome. On 16 January the airfield at Pitomnik had to be abandoned as a landing ground due to intense Soviet gunfire. Desperate Sixth Army officials in the Stalingrad pocket reported that the airfield at Gumrak could still be used but that the snow was much too deep and the troops too weakened by hunger to pack it down to withstand the impact of a landing aircraft. Nevertheless, an attempt to land was made by 20 Ju 52/3ms flown by the best and most experienced crews, but half the formation crashed and their wreckage blocked the airfield so that further attempts were extremely risky. During the first 22 days of 1943, an average of only 123 tons per day reached the encircled army. Then, on 23 January, Gumrak, the sole remaining operational airfield within the pocket, was overrun. The transports could not now land anywhere within the pocket. Supplies now had to be dropped by parachute or simply pushed out of the aircraft doors, and it was no longer possible to evacuate the wounded.

By 26 January, the pocket at Stalingrad had been split into two. At the end of the month, supplies were still being dropped into the two pockets but the troops were so weak they were unable even to retrieve the supply containers, many of which landed in deep snow, in the ruins, or were blown into Russian hands. The transports continued to brave the fire put up by a corridor of anti-aircraft guns positioned along their route, but

BELOW: When it proved impossible for transport aircraft to land, supplies were dropped to German strongpoints by parachute. Here, a supply container is being recovered, but by the end of January 1943, German troops at Stalingrad were too weak to retrieve them.

losses were extremely high. During the period between 24 November 1942 and 2 February 1943, when the last starved and exhausted remnants of Sixth Army finally surrendered, the *Luftwaffe* had carried or dropped a total of 8,250 tons of rations, fuel and ammunition to the encircled troops, an average of 114.58 tons per day, and had evacuated almost 30,000 wounded. But 488 aircraft, the equivalent of five *Geschwader*, had been lost and some 1,000 flying personnel had been killed or reported missing, the majority experienced and irreplaceable crews. As for the troops in Stalingrad, 91,000 surrendered but as a result of malnutrition and overwork in Soviet labour camps, only some 4,000 survived. The last survivors did not return to Germany until 1955.

RIGHT: Supplies being unloaded from a winter camouflaged Ju 52/3m, almost certainly of FFS (C) 16, one of the many schools units which were called up for operational duties and formed into provisional transport units.

BELOW: A feature observed on many Ju 52/3ms, such as this example of KGrzbV 102, was the black undersurfaces.

RIGHT: By January 1943, the only bases from which the transports could still operate to supply Stalingrad were at Novercherkassk in southern Russia and Salsk in the northern Caucasus. Apart from aircraft wear and tear, operations were disrupted by Soviet ground and air attacks and transport units at these airfields had to contend also with snowstorms. These grounded aircraft for days at a time, after which machines had to be dug out before sorties could be resumed.

LEFT: A similar scene on an airfield employed by I./KGzbV 1.

RIGHT: Troops working dangerously close to the revolving propellers of a Ju 52/3m while busily engaged in clearing a path for the machine to taxi.

ABOVE AND BELOW: The Luftwaffe made enormous efforts to supply Sixth Army at Stalingrad and even the Fw 200s of 1. and 3./KG 40 were withdrawn from Bordeaux and Trondheim and brought in to assist, operating under the temporary designation KGrzbV 200. Subsequently, these aircraft were withheld in the East to supply the Kuban bridgehead, and although they performed very well, the aircraft had originally been designed as an airliner and the structure of the military version was weak. Although the Fw 200 C-3 was strengthened, it continued to experience trouble with the rear spar which invariably failed when the aircraft was landing, the result being a broken back, as seen here. This machine is an Fw 200 C-3, W.Nr. 0025, and although it retained the operational code F8+FW of an aircraft of 12. Staffel, IV. Gruppe, it had at some time evidently been transferred to I. Gruppe and carried the badge of I./KG 40 on the nose.

ABOVE AND BELOW: A Ju 290 A-1 photographed in late 1942 during the Stalingrad airlift. Together with a Staffel of Ju 90s and the Fw 200s of I./KG 40, some of which may also be seen in these photographs, the Ju 290s were amalgamated throughout the airlift into a single, mixed Gruppe known as KGrzbV 200.

A snow camouflaged Ju 52/3m of 15./KGzbV 1 on the Eastern Front, almost certainly during the winter of 1942/1943. The use of tactical codes on the tail served no useful purpose after May 1943, when the Transport units were re-organised, but continued in limited use.

Junkers Ju 52/3m 1Z+FZ of 15./KGzbV 1, Russia, Winter 1942/43

This winter camouflaged aircraft had been modified with a de-icing system in which the exhausts from the wing engines ducted into non-corrugated covers positioned over the leading edge of each wing. The tail markings consisted of a tactical code, D3F, on the rudder and the last two letters of the operational code at the top of the fin. Note that the second, fourth, fifth and sixth side windows, together with the door window, had been blanked off and that the inside rear of the canopy had also been modified. Only the lower part of the yellow fuselage band was retained, the upper section having been overpainted with the white winter finish. The unit code 1Z on the fuselage was in small lettering.

Awards and Decorations for *Lufttransport* Aircrew

In 1941 it was realised that a decoration for bravery was required by the armed forces to bridge the gap that existed between the Iron Cross, First Class, and the Knight's Cross. The result was the institution of the German Cross in Gold, but although introduced in September 1941, it was first awarded to *Lufttransport* personnel only on 31 January 1942 when 20 flying personnel, all of III./KG zbV 1, were finally rewarded with the presentation of this decoration. These included three of the *Gruppe's* four *Staffelkapitäne* – Hptm. Gerhard Dudeck, Hptm. Fridolin Fath and Oblt. Hans Garms – with the remainder being awarded to various other aircrew from *Feldwebel* to *Oberleutnant*.

BELOW: The *Frontflugspange für Transport- und Luftlandeflieger*. This example is in silver and was normally awarded after 60 operational flights. (Author's collection)

Meanwhile, in November 1941, another award had been instituted to acknowledge the number of operational flights being made by the flying personnel of the various arms of the *Luftwaffe*. This was the *Frontflugspange*, usually translated as a War Flight Clasp or Operational Flight Clasp, and was awarded in three grades. Although the number of flights required to qualify for each grade varied during the war, as a rough guide, 20 operational flights were rewarded with the *Fronflugspange* in bronze, 60 were required for silver and 110 for gold. The design for the *Transport- und Luftlandeflieger* featured the *Luftwaffe* eagle and swastika enclosed within a wreath, to either side of which was a spray of oak leaves.

ABOVE: Despite its name, the German Cross in Gold contained no precious metals and the description referred to the colour of the wreath surrounding the swastika. The German Cross was a prestigious award and, in Gold, was introduced to recognise the bravery of the fighting troops. An almost identical award, the German Cross in Silver, had a silver wreath and was awarded to non-combat personnel for such feats as significant acts of leadership that furthered the war effort, rather than for acts of bravery. (Author's collection)

Towards the end of 1942, *Reichsmarschall* Göring suggested that a number of men on active service with the transport units should be decorated with the *Ritterkreuz*. Such a move was certainly long overdue, as the transport units had already shown themselves to be of particular value and the importance of their work could scarcely be overestimated. Yet prior to December 1942, when eleven Knight's Crosses were awarded to operational *Lufttransport* personnel, the only other similar awards to transport personnel since its institution in September 1939 had been in February 1942 to *Oberst* Friedrich-Wilhelm Morzik, the then *Kommodore* of KGzvBV 1, and *Oblt.* Lorenz Möller of II./KGzvB 1. Thus Göring's belated decision to decorate, for the first time, a group of men from the air transport units, was a clear recognition of his dependence on them and the ever-increasing difficulties under which they were obliged to operate.

RIGHT: The Knight's Cross of the Iron Cross was, until the introduction of further grades, the highest of Germany's awards for gallantry. As far as is known, only 44 Transportflieger were awarded the *Ritterkreuz*, and non received the next highest grade, the Knight's Cross of the Iron Cross with Oak Leaves, or above. (Author's Collection)

RIGHT: Fw. Herbert König, a pilot with 12./KGzvB 1, was awarded the German Cross in Gold in January 1942. In this photograph, König is shown wearing the decoration on his right breast pocket, while above the opposite breast pocket is the Operational Flight Clasp, or *Frontflugspange*. When König qualified for this latter award, a specific design for the Transport und Luftlandeflieger had not been instituted and aircrew were instead awarded the *Frontflugspange für Kampf- und Sturzkampfflieger*, as seen here. Although this situation was rectified with the institution in November 1941 of a clasp for transport aircrew, the older award continued to be worn. The other badges and medals being worn are the EK I, Pilot's Badge, and Wound Badge in Black.

Ritterkreuzträger of the *Transporter* Units, 1940-1945

Name and Rank	Unit	Date of Award	Remarks
Oberst Dipl. Ing. Gerhard Conrad	KGzbV 2	24 April 1940	*Kommodore*
Hptm. Peter Ingenhoven	KGrzbv 103	11 May 1940	† 1 February 1942
Obstlt. Gustav Wilke	KGrzbV 12	24 May 1940	*Kommandeur*
Oblt. Lorenz Möller	II./KGzbV 1	4 February 1942	
Oberst Friedrich-Wilhelm Morzik	KGzbV 1/*Luftwaffentransportführer Ost*	16 April 1942	*Kommodore/Luftwaffentransportführer*
Major Markus Zeidler	KGrzbV 600	9 May 1942	
Major Fridolin Fath	IV./KGzbV 1	24 December 1942	*Kommandeur*, joined post-war *Bundesluftwaffe*
Fw. Günther Frenzel	11./KGzbV 1	24 December 1942	
Oberst Theodor Beckmann	IV./KGzbV 1	24 December 1942	*Kommandeur*
Oblt. Josef Belz	KGrzbV 500	24 December 1942	† 2 November 1944
Fw. Oskar Kräussel	KGrzbV 172	24 December 1942	† 6 March 1943
Ofw. Walter Dominikus	IV./KGzbV 1	24 December 1942	
Major Erich Zähr	KGrzbV 172	24 December 1942	
Oblt. Walter Meltzer	III./KGzbV 1	24 December 1942	† 15 August 1943
Hptm. Ernst-Hermann Mersmann	1./KGrzbV 9	24 December 1942	
Lt. Alfred Köditz	I./KG 100	24 December 1942	Flew transport missions during Stalingrad airlift
Oberst Otto-Lutz Förster	*Luftwaffentransportführer Luftflotte* 4	24 December 1942	Also *Kommodore* KGzbV 1
Hptm. Kurt Geissler	*Lufttransport Gruppe* Don	24 January 1943	† 6 or 7 September 1943
Ofw. Hans Schoefbeck	1./KGrzbV 9	14 March 1943	
Obstlt. Ludwig Beckmann	KGrzbV 500	14 March 1943	*Kommodore*
Oblt. Martin Körner	4./KGrzbV 9	14 March 1943	
Hptm. Kurt Benz	KGrzbV 500	24 April 1943	Joined post-war *Bundesluftwaffe*
Hptm. Edgar Schwaneberg	2./TG 3	26 March 1944	Joined post-war *Bundesluftwaffe*
Oblt. Walter Bordelle	5./TG 2	26 March 1944	
Major Josef Grons	I./TG 2	20 April 1944	
Hptm. Heinz Hinkes	IV./TG 4	20 April 1944	
Lt. Hans-Joachim Valet	3./TG 2	20 April 1944	
Ofw. Günther Schmitz	10./TG 2	20 April 1944	
Hptm. Gerhard Dudeck	III./TG 2	9 June 1944	*Kommandeur*, joined post-war *Bundesluftwaffe*
Ofw. Max Hengst	2./TG 3	9 June 1944	
Major Walter Hornung	III./TG 2	9 June 1944	*Kommandeur*, joined post-war *Bundesluftwaffe*
Ofw. Karl Kern	6./TG 3	9 June 1944	
Fahnenjunker Fw. Fritz Kolb*	5./TG 3	9 June 1944	
Ofw. Herbert König	12./TG 1	9 June 1944	
Oberst Adolf Jäckel	TG 1	19 August 1944	*Kommodore*, † February 1945
Ofw. Hermann Schliermann	5./TG 4	31 October 1944	
Fahnenjunker Fw. Alfred Kummer	11./TG 2	18 November 1944	
Ofw. Paul Zebhauser	16./TG 1	18 November 1944	
Ofw. Hans Rötsche	16./TG 1	?? December 1944	Joined post-war *Bundesluftwaffe*
Ofw. Leonhard Ziehr	16./TG 1	8 January 1945	
Ofw. Erich Jaschinski	I./TG 3	9 February 1945	
Oblt. Hans-Heinrich v. zur Mühlen	2./TG 3	28 February 1945	
Lt. Wilhelm Messer	11./TG 2	12 March 1945	

* Kolb was an officer cadet when nominated for the RK but was a *Leutnant* when it was presented.

Generalleutnant Dipl. Ing. Gerhard Conrad, was awarded the Ritterkreuz while an Oberst and Kommodore of KGzbV 2 on 24 April 1940.

Little is known of Hptm. Peter Ingenhoven, one of the earliest Knight's Cross holders of the Transportflieger. At the time of the award, in May 1940, it is thought that he was serving with KGrzbV 103. Hptm. Ingenhoven was killed on 1 February 1942.

Major Markus Zeidler, received the German Cross in Gold while with KGrzbV 9 in February 1942. By May, when he was awarded the Knight's Cross, he had transferred to KGrzbV 600 as Kommandeur.

Fridolin Fath was a Hauptmann and a Staffelkapitän with IV./KGzbV 1 when awarded the German Cross in Gold in May 1942. In December of the same year, by which time he was a Major, he was awarded the Ritterkreuz. Major Fath survived the war and joined the post-war Bundesluftwaffe.

Fw. Günther Frenzel of 11./KGzbV 1.

As an Oberstleutnant and Kommandeur of KGrzbV 'Posen', Theodor Beckmann was awarded the German Cross in Gold on 4 May 1942. His Knight's Cross was awarded on 24 December 1942 when he was an Oberst and Kommandeur of IV./KGzbV 1.

Oblt. Josef Belz was flying with KGrzbV 500 when he was awarded the Ritterkreuz on 24 December 1942 but was killed on 2 November 1944.

Ofw. Walter Dominikus, a Ritterkreuzträger of IV./KGzbV 1.

Oblt. Walter Meltzer of III./KGzbV 1 received two high decorations in 1942. In January he was awarded the German Cross in Gold, and in December the Knight's Cross. He was killed on 15 August the following year.

Hptm. Ernst-Hermann Mersmann, a pilot and a Staffelkapitän with KGrzbV 9, was awarded the German Cross in Gold in February 1942 and the Knight's Cross in December.

Although not a Transportflieger, Lt. Alfred Köditz flew transport missions during the Stalingrad airlift with the bomber unit I./KG 100 and was engaged in transport missions when awarded the Ritterkreuz on 24 December 1942.

Oberst Otto-Lutz Förster received the Ritterkreuz on 24 December 1942 when he was Kommodore of KGzbV 1 and Luftwaffentransportführer with Luftflotte 4.

Hptm. Kurt Geissler was awarded the Ritterkreuz while serving with Luftwaffen Kommando Don in January 1943 but was killed in the following September.

Hans Schoefbeck was an Oberfeldwebel serving with 1./KGrzbV 9 when he was awarded the Knight's Cross on 14 March 1943. Later, after being commissioned as a Leutnant, he was awarded the German Cross in Gold while serving with I./TG 3 in January 1945.

Ludwig Beckmann was the Kommandeur of KGrzbV 500 in the Mediterranean and on the Eastern Front in 1942, receiving the German Cross in Gold on 16 July 1942. He was still leading the Gruppe on 14 March 1943 when, with the rank of Oberstleutnant, he was awarded the Knight's Cross.

Oblt. Martin Körner of 4./KGrzbV 9 was awarded the German Cross in January 1943 and the Knight's Cross in March.

Hptm. Kurt Benz was flying as Staffelkapitän with KGrzbV 500 when he was awarded the German Gross in Gold on 21 May 1942 and the Knight's Cross on 24 April 1943.

Ofw. Max Hengst of 2./TG 3 was awarded the Ritterkreuz on 9 June 1944.

Ofw. Paul Zebhauser of 16./KGzbV 1 received the German Cross in Gold in January 1943. He remained with 16./KGzbV 1 when it was redesignated 16./TG 1 in May 1943, and he still held the rank of Oberfeldwebel when he received the Ritterkreuz in November 1944.

Ofw. Hans Rötsche was flying with 16./TG 1 when he was awarded the Ritterkreuz in December 1944.

As a Feldwebel pilot with 1./KGrzbV 'Posen', Leonhard Ziehr received the German Cross in Gold in June 1942. Later, as an Oberfeldwebel, he was flying with 16./TG 1 when he was awarded the Knight's Cross on 8 January 1945.

A Ju 52/3m g7e showing the large loading door on the starboard side of the fuselage which was a feature of this sub-type. Although the unit code ahead of the cross has been obscured by the temporary winter finish, the tail markings and the Staffel letter Q indicate that this machine belonged to 16./KGzbV 1. The upper gun on this aircraft is an MG 131 and required a different mounting to the usual MG 15.

Junkers Ju 52/3m g7e of 16./KGzbV 1, Russia, Winter 1942/1943

The temporary white camouflage on this aircraft had been applied by brush, as opposed to being sprayed on, and the resulting patchy nature of the finish allowed areas of the original green 70/71 to show faintly through. Although the full operational code on this machine was 1Z+FQ, only the last two letters have been retained and appear again on the fin. Although lightly overpainted, the tactical code D4F on the rudder is still visible and was probably later repainted in black to make it more conspicuous. Note the de-icing equipment for the leading edges of the wings and the areas under the wings where a Stammkennzeichen had been overpainted.

A lone member of the ground staff clearing snow from the wings of a Ju 52/3m of IV./KGzbV 1, probably photographed during the winter of 1942/1943. Note the covered windows and that a narrow border of the original 70/71 has been left around the letter L on the fuselage in order to make it conspicuous against the camouflage colour. The 1 in the tactical code on the rudder indicates the first Staffel of the IV. Gruppe, i.e. the 13. Staffel in the Geschwader. A full explanation of the origins and use of these tactical codes appears in Volume 2.

Junkers Ju 52/3m of 13./KGzbV 1, Russia, Winter 1942/1943

Although rather patchy, the recently brush-applied snow finish on this aircraft still appears lighter than the white of the fuselage Balkenkreuz due to the dirt and exhaust staining which had previously accumulated on the fuselage. Only the last two letters of the operational code 1Z+LX have been retained on the fuselage, and these are repeated on top of the fin in black, evidently applied over the snow finish. The dark borders of 70 and 71 around all other markings show where care had been taken to avoid overpainting them.

The Air Supply of Seventeenth Army

The battles to encircle Stalingrad had placed the Soviets in an ideal position to exploit their winter successes and launch further attacks, so threatening to trap First *Panzer* Army and Seventeenth Army – which together comprised Kleist's Army Group A – in the Caucasus. First *Panzer* Army managed to escape through a narrow corridor to Rostov before this was sealed on 14 February 1943, but Seventeenth Army was trapped and forced to retreat to the Taman Peninsula where, with the Sea of Azov at its back, it was completely isolated. Although a retreat across the Kerch Straits was possible, Hitler ordered that this area, which became known as the Kuban bridgehead, was to be held.

RIGHT: A Ju 52/3m over the Sea of Azov, probably during a mission to Army Group A in the Caucasus in early 1943. The tactical code letter Z on the rudder had been assigned to III./KGzbV 1 in December 1942. As this Gruppe was in the Mediterranean theatre in early 1943, the aircraft was probably withdrawn for overhaul and then sent to another unit in the East, still with its tactical code.

BELOW: The Soviet Union was not a signatory to the Geneva Convention and did not recognise the status of Red Cross aircraft, hence the MG 15 mounted for defence on this ambulance aircraft. It may be noted, however, that Luftwaffe ambulance Ju 52/3ms operating in the East were sometimes employed to carry ammunition, and five such aircraft were recorded as taking fuel to the North Caucasus in February 1943.

Despite the bitter fighting, Seventeenth Army was in a strong position and although supplies were available for the immediate future, it was decided to preserve the Army's fighting power by the use of air transport which would fly in the necessary supplies of food, fuel and ammunition. Wounded and other personnel were to be evacuated during return flights, and some transport aircraft were also to fly out valuable supplies of copper to Zaporozhye. However, the Stalingrad operation had all but exhausted the transport units in the East and reduced their effectiveness to a minimum. In addition, when *Luftflotte* 4 turned to the new task, it found the available airfields too few and inadequately prepared, while the units themselves, still based on the airfields they had used in the final stage of the Stalingrad operation, were in the wrong place for working in the Caucasus. There was thus a delay before the transport force could be fully committed and from 29 January to 3 February, air supply and evacuation was carried out mainly by a handful of ambulance Ju 52/3ms.

Another effect of the high losses sustained during the Stalingrad operation was that the transport units were reinforced with what proved to be the largest force of gliders ever assembled for a single *Luftwaffe* operation. Due to a lack of airfield facilities, however, the glider *Gruppen* were unable to commence operations until late January; transport work was meanwhile carried out by the Fw 200s of KGrzbV 200.

At the end of the Stalingrad operation, KGrzbV 200 had been ordered to Berlin–Staaken for refitting before returning to its parent unit, KG 40, on the Atlantic Front, but these orders were temporarily countermanded by the instruction to commit every available aircraft to the supply of the Kuban bridgehead. The Fw 200 operations commenced on 4 February with the aircraft performing extremely well. A high degree of operational readiness was achieved, largely due to the arrival from France of a trainload of technical personnel and special equipment. These had, in fact, been requested weeks previously during the Stalingrad operation, but despite being assigned top priority, their train had been shunted from station to station in Poland in order to let the *Waffen-SS* Division *'Das Reich'* through to the front. KGrzbV 200's supply missions continued until 13 February, during which time the aircraft carried all manner of supplies and ammunition to various airfields in the Kuban bridgehead, and flew out wounded on the return flights [8].

8. KGvbV 200 withdrew to Berlin-Staaken on 22 February 1943 and was disbanded in March. The 1. and 3./KG 40 were then united to form a new 8./KG 40.

Meanwhile, the Ju 52/3m units from the Stalingrad area had been refitted and reorganised, so that it was possible to assemble five *Gruppen* (KGrzbV 9, KGrzbV 50, KGrzbV 102, I./KGzbV 172 and KGrzbV 500) on airfields in the Crimea. Considerable difficulty was encountered in finding suitable airfields in the Kuban, however, and the weather was too cold to permit the use of Ju 52/3m floatplanes owing to ice in the Kerch Straits. The ice, however, was not thick enough for use as a runway, nor was it cold enough to freeze permanently the short, extemporised runways that were liable to become softened during the day. As landings, therefore, were only possible during the morning and late afternoon, other supplies delivered during the rest of the day had to be air dropped.

By the end of January, a number of tow aircraft and gliders from I., II. and III./LLG 1 had joined operations equipped with a mixture of Do 17, He 45, He 46 and Hs 126 towing aircraft and DFS 230 gliders. These had been especially requested by *Generalfeldmarschall* Milch during the Stalingrad airlift, but all the take-off and landing bases lacked the necessary facilities, which prevented the gliders from being employed during the Stalingrad operation. They had, however, been transferred to the area, and once in position on the Crimean airfields of Bagerowo and Kerch IV, their use during the supply of Seventeenth Army greatly increased transport capacity.

Apart from the three *Gruppen* of LLG 1, the force included I./LLG 2 with He 111s and Go 242s, and a special *Gruppe* of Me 321 Gliders and He 111 Z tugs making their operational debut. The first limited air-supply operations to the Kuban bridgehead began at the end of January, and although these units had no winter ground equipment and the glider units had no experience of winter conditions in the East, they made a significant contribution to the undertaking. Supplies included ammunition, food, and especially fodder for the horses of the mountain troops in the encircled area. Their effort was also notable as heavy traffic at the take-off and landing areas called for a great deal of care and skill on the part of the tug and glider pilots.

Operations were made even more difficult as the landing grounds thawed out, though the higher temperatures melted the ice in the Kerch Straits. Ju 52/3ms from all fronts were therefore hastily converted to floatplanes, while 22 Do 24 rescue aircraft were speedily brought up from the Mediterranean and joined operations in March. Early that same month He 111s, which had been used as bombers since the Stalingrad operation, came into use as transports and also acted as tugs for the Go 242s of LLG 2. The final thaw arrived around the end of March, and by early April it became possible to supply Seventeenth Army by sea, allowing the bulk of the exhausted air transport units to withdraw to Germany and refit. Only KGrzbV 102 and the He 111 units remained to complete air transport tasks in the area.

LEFT: The five-engined He 111 Z – the Z indicating Zwilling, meaning Twin – was specifically designed to tow heavy gliders. This rare photograph shows an He 111 Z photographed at Prossnitz (Prostejov) in Czechoslovakia in early 1943, at which time it was probably en route from its base in France bound for the Eastern Front where it was intended it should take part in the Stalingrad airlift. The He 111 Zs arrived too late, however, and first saw practical use towing Me 321 gliders during the operation to supply the Kuban bridgehead. This example carried the Stamm-kennzeichen DG+DY.

1939-1943

ABOVE: During the operation to supply the Kuban bridgehead, the unusual He 111 Zs operated from Bagerowo in the Crimea under Luftlandegeschwader 1. These aircraft and their towed gliders succeeded in evacuating large numbers of wounded; an He 111 Z with an Me 321 in tow was capable of carrying up to 130 casualties in one lift.

LEFT: Wearing its snow camouflage, this He 111 Z, shown on a muddy airfield in the thaw in the spring of 1943, has been fitted with external long-range tanks.

1939-1943

TOP AND ABOVE:
Another example of the
He 111 Z. The machine
shown here is believed
to have belonged to
Schleppgruppe 2.

LEFT: Unloading
supplies from a Go 242
glider at Stalino in
southern Russia in
early 1943.

1939-1943

ABOVE: A winter-camouflaged Me 321 of GS-Kdo.2 (Grossraumlastenseglerkommando 2) at Kerch in the Crimea at the end of January 1943. An He 111 Z, used for towing the Me 321 gliders, may be seen in the background.

BELOW: A closer view of the same Me 321, coded W6+SW. A large proportion of the loads carried from the Crimea into the Kuban consisted of hay and straw to feed the horses employed by the Gebirgsjäger.

Fluggruppe — IV/43 —

Gr. Stab der Walfische

ABOVE: Of the approximately 200 Me 321s completed, some saw service with GS 1, GS 2, GS 4 and GS 22. It is thought that these Kommandos later formed the Grossraumlastensegler Gruppe and that this may have been known as the Walfisch, or Whale, Gruppe. This photograph shows the emblem of the Stab of the Walfisch Gruppe painted on a signpost pointing the way to the office of the Gruppen Stab at Obertraubling where the Gruppe was first formed.

A Ju 52/3m of 14./KGzbV 1 in Russia with yellow theatre markings and white winter camouflage. Note that the door has been removed and a member of the crew is about to release supplies. The tactical code on the fuselage, 14./KGzbV 1 can be established from the tactical code on the rudder and the Staffel letter Y on the fuselage, while the presentation of the Geschwader code 1Z, shown in small characters ahead of the fuselage Balkenkreuz, together with the tactical code and winter finish, all suggest early 1943. The upper gun position consisted of an MG 131 instead of an MG 15.

Junkers Ju 52/3m 1Z+KY of 14./KGzbV 1, Russia, early 1943

This winter camouflaged aircraft has a partly overpainted fuselage band and an interesting combination of tail markings comprising a tactical code on the rudder and the last two letters of the operational code on the top of the fin. Note the unusual style of the letter Y and that the temporary white finish has been applied quite thinly, allowing the original green to show through. The original 70/71 finish can be seen in the small rectangle on the fuselage containing the unit code and around the letter K, and also in the two stripes at the top of the fin.

1939-1943

ABOVE: The view from the cockpit of an Me 321 under tow by an He 111 Z of the so-called Walfisch Gruppe.

LEFT: Luftlandegeschwader 1 used a great variety of aircraft as glider tugs, including these Avia B.534 biplanes of Czech origin shown towing DFS 230 gliders.

BELOW: This Me 323 D-1, W.Nr. 1220, was allocated the Stammkennzeichen DT+IT and carried the name 'Himmelslaus' (Heavenly Louse) in small white lettering on the nose as a pun on the aircraft's size. The machine was flown by II./KGzbV 323 and is known to have made a crash-landing at Lemberg (Lvov) in southern Poland on 15 February 1943. Subsequently, when II./KGzbV was redesignated II./TG 5 in May 1943, the aircraft remained with the Gruppe and is shown in this photograph at Kharkov in the summer of 1943. Note the areas of red dope around the undercarriage where the fuselage has been recently repaired.

ABOVE: Ju 52 of LLG 1 loading Fallschirmjäger in the Crimea in 1943. Following the near disaster in Crete in 1941, the parachute troops were largely committed to ground fighting.

BELOW AND INSET: Two aircraft of Schleppgruppe 1 at Kharkov in 1943. The DFS 230 glider (*BELOW*), apparently camouflaged in 02 and 71, has the fuselage code F7+7, while the Hs 126 B-1 tug in the background was marked F7+ZP. Another view (*INSET*) of the Hs 126 F7+ZP showing the small unit code in white and the yellow wingtips and lower cowling.

Air Transport in the Mediterranean

The capture of Crete in early June 1941 had brought about the transfer of X. *Fliegerkorps* from Sicily to Greece and the centre of air transport to German forces in Africa shifted from Italy to Athens, with Crete as the obvious transit station between Europe and Africa.

At this time, the German air transport fleet, which had already suffered considerable losses in the Crete campaign, was further reduced as units were recalled to Germany for the invasion of Russia. By the end of June, of the units which had been in the area at the time the island was captured, only III./KGzbV 1 and KGrzbV 9 remained and the Mediterranean theatre was thus left very short of air

LEFT: British Commonwealth troops examining the remains of a Ju 52/3m, almost certainly of KGrzbV 9 in North Africa. This machine, probably captured during the 'Battleaxe' offensive of June 1941, still carries the yellow cowling and rudder markings applied during the Balkans and Crete campaigns of April and May 1941.

transport at a time when it was about to need it greatly. From this time onwards, there was never a really adequate supply of transport aircraft and crews for the Russian and Mediterranean fronts and generally the Mediterranean forces suffered from the greater urgency of supplies to Russia. Servicing facilities in Italy, Greece and Africa were as yet inadequate and even after the Italians had been prevailed upon to release a *Staffel* of SM.82s to operate with III./KGzbV 1 at Lecce, there was still a shortage of aircraft to deal with the demand for supplies.

In November 1941, with the opening in Libya of the British 'Crusader' offensive, which was intended to relieve Tobruk and destroy the *Afrika Korps*, air transport became absolutely essential and, since no units could be spared from the Russian Front, KGrzbV 300 was formed from the C-*Schule* at Hörsching and sent to Athens-Tatoi. Further reinforcements arrived in December, when KGrzbV 400 and KGrzbV 500 were set up from the *Blindflugschulen* (Blind Flying Schools) at Brandis, Wesendorf, Radom and Neuburg, while a *Staffel* of I./LLG 1 was moved from Crete to Eleusis in mainland Greece. This reinforcement was of short duration, however, as KGrzbV 9 and KGrzbV 500 were withdrawn in January to meet supply needs in Russia. Missions to Africa were therefore continued by III./KGzbV 1 and KGrzbV 400 alone, but these two units were unable to cope with the total requirements. Moreover, KGrzbV 400, which had already lost 11 machines, continued to lose aircraft and personnel as a result of enemy action and the young and inexperienced crews were unable to meet the demands made of them.

From mid-December, supplies consisted mainly of fuel for the retreating *Afrika Korps*, but when KGrzbV 400 was moved to Tripoli on 26 December to ferry troops from Trapani to Africa, petrol cargoes

BELOW: Ju 52s taking off from Sicily on a supply mission to North Africa.

became fairly rare and were only undertaken when absolutely necessary. Such missions were favoured by an almost continuous haze and rough seas which reduced the likelihood of encountering Malta-based fighters. The British fighters were, in any case, wary of approaching if the transports immediately opened fire and, using this experience, the transports adopted defensive measures. When attacked, each transport reduced height until it was just above the water to prevent the fighters attacking from below, and any infantry aboard were ordered to open fire through the aircraft's windows. The combination of low flying and the concentrated firepower of the entire formation of perhaps 50 or 60 aircraft usually constituted a successful defence, and after two British fighters were shot down by transport aircraft on 8 January, the confidence of the transport crews increased and the fighters seemed even more willing to leave them alone. Indeed, for several weeks no fighters were sighted at all. Thus a false sense of security prevailed and the virtually unescorted transports began flying the same course at the same time day after day.

Meanwhile, the ground situation had improved to such an extent that the *Afrika Korps* was advancing again. At first, all available transports were ordered to fly petrol up to the *Luftwaffe's* forward fighter and dive-bomber units, the aircraft making local flights from Crete to Benghazi and Derna. Later, however, troop replacements became first priority loads, KGrzbV 400 being particularly active in bringing troops from Italy via Greece to Derna.

ABOVE: The greatest problem in the desert war, and one which proved crucial, was that of supply. Ultimately, the victor would be the side which received the best supplies in the shortest time, and with few harbours at which distribution depots could be established, all such essentials as petrol, water, ammunition and food, as well as men and equipment, had to be transported along long lines of communication. The value to the Germans of the Ju 52/3m was therefore considerable, and even a minor accident, such as has occurred here with this aircraft of KGzbV 1, was of some significance if it rendered the machine unserviceable.

ABOVE: The operational code 1Z+LK on the fuselage of this Ju 52/3m captured at Tmimi in January 1942 confirms that it had originally served with 2./KGzbV 1, and the appropriate I. Gruppe badge, a chess-piece knight, appeared on the forward fuselage. However, the rudder letter P indicates the aircraft had been temporarily assigned to KGrzbV 400 which used this letter between December 1941 and November 1942. The significance of the 7 is unclear but may be an alternative to using a letter to identify the individual aircraft.

RIGHT: This machine at Maleme also belonged to either KGzbV 1 or KGrzbV 400. Although the colours in the photograph have faded, the yellow undersides of the engine cowlings may still be recognised.

THIS PAGE: After the German conquest of Crete in 1941, the island became an important staging post for aircraft flying from Italy and Greece with supplies for Rommel's Afrika Korps. It was also built up to form an island fortress, thus imposing a further task on the transport units. In the first half of 1942, the Luftwaffe had established air supremacy over the central Mediterranean so that, during the battles in Libya, supply flights were largely uninterrupted and the airfields in Crete offered the shortest flying distances to the Benghazi – Derna area. These photographs, with the Spatha Peninsula in the background, show Ju 52/3ms of KGzbV 1 and KGrzbV 400 at Maleme between flights to North Africa. Note the contrast *(RIGHT)* between the black of the Balkenkreuz and the green 70 and 71 camouflage paints which shows the extent to which these colours faded and weathered.

As the number of Ju 52/3ms flying daily into Derna began to increase, the RAF started to keep watch, and on 12 May a force of five Beaufighters escorted by nine P-40s intercepted a formation of 13 Ju 52/3ms from III./KGzbV 1, escorted by a single Bf 110, 160 km off Derna. The Ju 52/3ms were carrying troops, and the Beaufighter leader later reported that a "multitude of tommy guns" was pushed through the windows as the Allied aircraft prepared for their first attack. The fight was witnessed in appalling detail by the fighter pilots. Eight Ju 52/3ms went down in flames over the water and 175 men were lost. Some of the troops were seen to jump or fall from the blazing aircraft before they hit the sea, while others perished as they struggled from the ditched machines into patches of burning oil. The Beaufighter leader later reported that the scene, with columns of smoke rising from the sea almost to the horizon, resembled a convoy of ships in line astern. Despite heavy damage, the remaining five Ju 52/3ms managed to reach the beach or the airfield at Derna but all had to be written off. The British lost a single Beaufighter which was shot down into the sea by *Uffz. Graf* von Polier, a gunner aboard one of the Ju 52/3ms. Subsequently, the flight routes and timings were constantly altered and each group of transports was escorted by four fighters.

Meanwhile, the operational readiness of the *Transportverbände* steadily declined. One *Gruppe* had only 18 to 20 serviceable machines available per day. Another had been assigned so many special missions that it had only ten or 12 aircraft available for supply missions to the front. Serviceable aircraft were flying as many as three missions per day, a total flight time of 12 hours, yet still the *Afrika Korps'* requirements could not be met.

The German capture of Tobruk on 21 June 1942 improved conditions for air transport as it brought in new landing grounds and provided a harbour suitable for the large, six-engined Bv 222 flying-boats of *Lufttransportstaffel See* 222 which operated under the control of KGzbV 2. At the beginning of July,

**Luftwaffe Air Transport Routes
in the Mediterranean**

Note: The routes
to Tunisia were
only opened after
8 November 1942.

ABOVE AND BELOW: Formations of Ju 52/3m transports over the Mediterranean. The photograph (*ABOVE*) was taken from a Bf 110 escort fighter.

ABOVE: For much of the African campaign, there were barely sufficient fighters available to escort the Ju 52/3ms and even when they could be spared, large formations were often escorted by only a few fighters and relied instead on their low altitude and combined defensive firepower. Note the barrel deflector in the foreground which prevented the gunner from damaging his own machine.

LEFT: Having crossed the Mediterranean and the North African coast, a formation of Ju 52/3ms from KGzbV 1 heads for its inland base.

however, there were still only 150 Ju 52/3ms in the Italy-Greece-Africa arena. The shortage of transport aircraft was so acute that the machines of *Transport Staffel/*II. *Fliegerkorps,* the *Nachrichten* (Signals) *Staffel* of X. *Fliegerkorps* and the ambulance Ju 52/3ms of *Sanitätsflugbereitschaft* 7 (Air Ambulance Duty Flight 7) were withdrawn from their parent units to reinforce the air transport effort. Further reinforcements took place in July with the arrival at Maleme of IV./KGzbV 1, KGrzbV 600 and KGrzbV 800. These three new *Gruppen,* together with the occasional assistance of LLG 1, increased the number of Ju 52/3ms to 300. That these aircraft were able to meet the *Afrika Korps'* demand for the daily transport of 1,000 troops and 25 tons of equipment was mainly due to the fact that it was now possible to store the daily fuel requirements of the transport units (approximately 360,000 litres) on Crete, and that servicing facilities had been set up in Africa.

At the end of August, however, the *Panzer Armee Afrika,* its offensive already losing momentum, ran critically short of fuel. Every available aircraft was needed to carry petrol from Crete to Tobruk and up to the frontal areas, but although Allied fighter units were weak after their long retreat, normal operational wear and tear had reduced *Luftwaffe* air transport to some 250 aircraft. Most *Gruppen* were operationally tired, and serviceability of machines and fitness of crews fell accordingly. Indeed, the shortage of crews became so acute that units were reduced to borrowing from each other in order to be able to fly whatever aircraft they had, and some partly trained bomber pilots were introduced as second pilots. Flying conditions were as difficult as ever, and fighter protection for the formations was inadequate. The obvious remedy was to fly transport operations at night, but such a decision was delayed as insufficient crews trained in blind flying were available. This shortage of night-flying crews was due, at least in part, to the *Luftwaffe's* policy of withdrawing instructor crews and aircraft from the C-*Schulen* and *Blindflugschulen* for the formation of fresh transport *Gruppen.*

The fuel shortage became very acute in September 1942 when fuel for mechanical transport and for Ju 52/3ms was in very short supply. Indeed, it was impossible to supply fuel for Rommel's tanks because there was insufficient petrol for the Ju 52/3ms. So bad was the situation that an auxiliary transit station for troop reinforcements was opened at Molaoi on the Greek mainland so that transport aircraft should have a shorter flight than if they had flown to and from Athens.

In October 1942, air transport in the Mediterranean area, pressed to the limit to supply Rommel, lacked aircraft, crews, fuel and fighter protection. It was, therefore, hardly in a position to redouble its efforts when the British Eighth Army went over to the offensive at El Alamein on 23 October 1942.

The first immediate effect of the El Alamein offensive on German air transport was the arrival of long-range Fw 200 aircraft at Lecce. These were soon reinforced by a *Staffel* of Ju 90 and Ju 290 aircraft at Lecce, which was also the base of the Savoia *Staffel* of KGzbV 1. A tremendous effort was made from Italy and Crete to supply Rommel's retreating forces, but while III./KGzbV 1 had received 20 new Ju 52/3ms at the end of October, a shortage of barrels arose which began to limit the supply effort.

On 3 November, the British broke through the Axis defences at El Alamein, forcing Rommel to withdraw his forces along the coast. Even as this was taking place, another plan was already in progress which the Allies intended would ensure the complete destruction of Axis forces in North Africa. At the end of October 1942, 39 vessels comprising the largest convoy ever to leave the United States, had sailed into the Atlantic Ocean and headed eastwards. Three thousand miles away, an even larger convoy of 81 vessels left the coast of Great Britain and sailed on a southerly course. Some days later, the two convoys met up, passed through the Straits of Gibraltar and sailed into the Mediterranean. At this point, the convoys were shadowed by *Luftwaffe* reconnaissance aircraft which reported their progress to OKW. There, it was assumed that the convoys, which remained largely beyond the range of German torpedo bombers, would continue eastwards to Malta or Egypt, but on 8 November, when the ships were off French Morocco and Algeria, they turned to starboard and headed for the coast. Operation 'Torch', the launching of a second front in North Africa, had begun.

Although the Allies had been unable to occupy Tunisia, their main objective in the first landings, it was thought that their forces would be able to advance into Tunisia before the Germans were able to react. In this estimation they were wrong, however, for although OKW had remained uncertain of the destination of the convoys after they had passed into the Mediterranean, their reaction to the landings was immediate. *Luftwaffe* dive-bomber and fighter units began arriving in Tunisia the following day, and by the 10th, the transport arm was active in ensuring the rapid transfer of German troops and equipment to establish a bridgehead in the area. Henceforth, with Rommel's army retreating westwards from Egypt, Tunisia became the focus of action, and the *Transporters* would face an even more difficult role in the months of fighting which lay ahead.

ABOVE: The six-engined Bv 222s of Lufttransportstaffel (See) 222 began flying troops and supplies across the Mediterranean to the Afrika Korps in the summer of 1942. Note the tactical code S4 and a white diagonal stripe on the rudder, the numeral indicating that this is the Bv 222 V4.

LEFT: Loading a Bv 222 with supplies.

1939-1943

BELOW: A busy airfield scene in North Africa as Ju 52/3ms – some still with the yellow markings applied for the Balkans campaign – fly in fuel. Often the only aircraft available to escort quite sizeable formations of transports was a pair of Bf 110 heavy fighters, two of which may also be seen in the photograph.

BELOW: Not to be confused with the Balkans markings, the yellow cowling surfaces on this Ju 52/5m in the Mediterranean in 1942 were a standard recognition aid on all German aircraft operating in this theatre.

ABOVE: Photographed at about the same time as the photograph (*ABOVE LEFT*) was the Ju 90 V-3. This aircraft, named 'Bayern', was completed in 1938 and was subsequently delivered to Deutsche Lufthansa, but shortly after the outbreak of war in 1939, it was taken over by the Luftwaffe. With its civilian markings hastily overpainted, the aircraft is shown here about to be refuelled in the Mediterranean theatre.

This Ju 52/3m g4e had the unit code IZ+AS of Stab, 8./KGzbV 1 on the fuselage but had been assigned to 3./KGrzbV 400, as confirmed by the tactical code on the rudder. KGrzbV 400 was formed in December 1941 and this aircraft is thought to have been photographed in February 1942 when the aircraft was operating in the Mediterranean flying supplies from Sicily to Rommel's forces in Libya.

Emblem of II./KGzbV 1

Junkers Ju 52/3m g4e 1Z+AS of 3./KGrzbV 400, February 1942

The operational code on the fuselage and under the wings indicates that this machine had earlier belonged to 8./KGzbV 1. However, when aircraft were commandeered and reassigned to other units by means of the so-called 'Kommandoweg', the original operational markings lost all meaning. It is therefore the tactical code P3F on the rudder of this machine, W.Nr. 6396, which indicates its unit, 3./KGrzbV 400, and that it was aircraft F. Note the handrail below the fuselage windows, the yellow areas on the engine cowlings, and the rear-view mirror mounted on the canopy. The camouflage was the standard 70/71/65 scheme.

ABOVE: This Ju 52/3m of KGzbV 1 photographed at Comiso in Sicily was probably operating in the support role when units of II. Fliegerkorps, and subsequently the whole of Luftflotte 2, were withdrawn from the Eastern Front in late 1941 to operate over Malta and North Africa. Note again the dark green or black undersurfaces and the individual aircraft letter K under the wing tip.

RIGHT: Although the DFS 230 was originally conceived as an assault glider, it was also used in the transport role, some examples being employed for that task in North Africa. The DFS 230 proved most useful in the transport of fuel and particularly of food, as the larger hold could be completely filled, especially if rusks and tinned goods were carried.

ABOVE: This earlier photograph of the same aircraft, taken at Hildesheim in 1941, shows the position of the Geschwader badge. In the cockpit is the then Kommodore of LLG 1, Obstlt. Gustav Wilke.

ABOVE: This wrecked Ju 52/3m abandoned in the Western Desert was originally assigned to the Kommodore of LLG 1, Oberstleutnant Gustav Wilke and later to his replacement, Generalmajor Rüdiger von Heyking. However, when KGzbV 400 was formed in December 1941, it took over a number of aircraft belonging to LLG 1, including this aircraft, W.Nr. 5055. Although the rudder was missing when photographed, it is known that this machine was marked with the tactical code P2D in yellow. All characters of the operational code H4+AA have been outlined in white, and as the aircraft had belonged to the Geschwader Stab when flying with LLG 1, the individual aircraft letter A was blue.

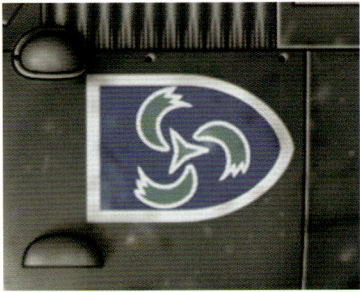

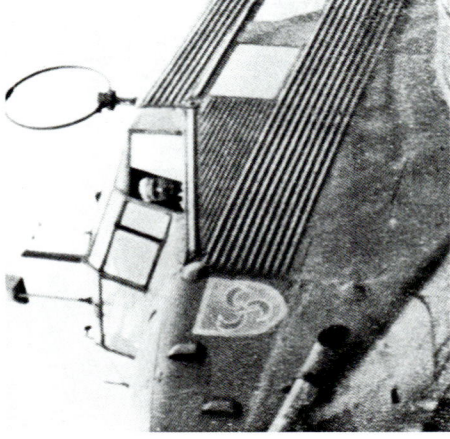

Emblem of LLG 1

Junkers Ju 52/3m of 2./KGzbV 400, Libya, 1942

During the Crete campaign in May 1941, this aircraft, W.Nr. 5055, was 60 per cent damaged at Maleme. Although damage to this extent would normally have resulted in the machine being written off, it was repaired and returned to service with LLG 1, but then passed to 2./KGrzbV 400. The badge and code of LLG 1 were retained but a new identity, P2D, was added as a tactical code with the D repeated to 2./KGrzbV 400. Note the faded 71/65 uppersurface camouflage, the mirror, and the large D/F loop.

ABOVE: Although carrying the operational code H4+CH of 1./LLG 1, it is thought that this is another aircraft which was operating with KGzbV 400 when photographed transporting supplies to North Africa in 1942. Note the so-called 'Condor-Haube' over the cockpit.

RIGHT: A Ju 52/3m being refuelled in Sicily. On some occasions during the airlift to Rommel's forces, it was found that in order to return to their bases, transports flying long-range supply missions had to be refuelled with half the quantity of petrol they had carried to the front in North Africa. Note the MG 15 mounted in the 'Condor-Haube' over the cockpit and the triangular tactical symbol on the fin and leading edge of the wing. This particular machine, G6+AP, belonged to KGrzbV 102 and was operating under the control of KGzbV 2.

1939-1943

ABOVE: A large number of Ju 52/3ms gathered at Comiso in Sicily. Unlike other Luftwaffe types operating in the Mediterranean, which adopted camouflage schemes usually involving tan or sand colouring, the Ju 52/3ms retained their dark green finish.

LEFT: One of the lesser-known units operating in the transport role in North Africa from late 1942 was Transport Staffel/II. Fliegerkorps based at Reggio, in southern Italy, and at Trapani in Sicily. This example, seen being examined by British Commonwealth personnel, was found abandoned in Libya. It had the operational code C3+KH on the fuselage and the Staffel badge on the nose. Note that this is not the same machine as shown on page 96.

LEFT: The wastage of transport aircraft in the Mediterranean and in Russia in 1941 and 1942 was such that production of Ju 52s could not keep pace with losses and the Luftwaffe was obliged to obtain SM.82s from the Italians. These aircraft proved most useful, possessing a greater speed than the Ju 52/3m and a larger cargo capacity. This photograph shows SM.82s of KGzbV 1 over the sand dunes of North Africa. The aircraft in the background, believed to have been coded IZ+MS, is in standard Luftwaffe camouflage, while the uppersurface camouflage on the wing of the machine from which the photograph was taken appears to be 70 or 71 with an overspray of a sand colour.

LEFT: A Ju 52 of 11./KGzbV 1 flying low over the sea. In preparation for Rommel's impending assault at Alam el Halfa in August 1942, an emergency airlift was organised to ferry petrol from Italy to Crete, from Crete to Tobruk, and from Tobruk to the front. Rommel's attack opened on 30 August but failed due as much to supply difficulties as to battlefield losses. After the battle, a shortage of tank and anti-tank ammunition became critical, followed by troop replacements, but so depleted had daily supplies become that priority had to be allocated to such basics as food and drinking water. Note the dark green or black undersurfaces on this machine and the individual aircraft letter K repeated in yellow on the leading edge of the wing.

RIGHT: Luftwaffe signals personnel of Flugkdo./Ln Regt.(mot) Afrika, having evidently disembarked from the Ju 52/3m in the background. This aircraft appears to be finished overall in green 71 and has the complete operational code repeated under the wings. The white fuselage band and wingtips are consistent with the Mediterranean theatre. The lightning bolt and small numeral on the rudder are yellow, but the circular device above it is white.

ABOVE: The airfield at Bari on the south-east coast of mainland Italy was used by aircraft making intermediate landings before flying on to North Africa. In this scene, probably in October 1942, Bf 109 G-2s of III./JG 27 share the airfield with two Ju 52/3ms. The aircraft furthest from the camera is W.Nr. 6007 and has the Stammkennzeichen TJ+HX, while that nearest the camera has the operational code H4+CR of 7./LLG 1. However, the tactical code P5 on the rudder of both machines indicates that they were serving with KGrzbV 400 when photographed. Note the small circle on the fin of H4+CR and the different positions of the white fuselage bands on each machine.

LEFT: In this formation of Ju 52/3ms, the tactical code on the rudder of the aircraft on the far left identifies the machine as belonging to III./KGzbV 1. Note the bleached appearance of the 70 and 71 on the wing in the foreground and that, unusually, the aircraft in the background, right, has retained its wheel spats.

1939-1943

ABOVE: A Ju 52/3m about to touch down during transport operations in the Mediterranean. This aircraft has the unit code 8T and the tactical rudder letter T of KGrzbV 800, which arrived in the Mediterranean theatre in July 1942. The tactical codes on the aircraft in the background indicate they were from III./KGzbV 1.

ABOVE: A crash-landed Ju 52/3m of KGrzbV 1 in North Africa. Spots, circles and triangles were often used as tactical symbols although their precise significance remains unknown.

RIGHT: This Ju 52/3m at Benina was probably destroyed in late 1942 and shows a yellow triangle outlined in white on the fin and a similar tactical marking on the top of the fuselage. Unfortunately, the significance of the triangle is not known, although the fact that this aircraft was normally flown by the Kapitän of a 7. Staffel may be relevant.

LEFT AND BELOW: Immediately before the battle of El Alamein in October 1942, RAF fighter activity in the area was low but increased after British and Commonwealth forces had achieved their breakthrough and moved their fighter squadrons further westwards. Although the situation over the sea remained quiet, fighter activity over the land steadily increased in intensity, disrupting Luftwaffe transport missions and causing constant losses. These photographs show *(LEFT)* RAF Beaufighters forcing down a Ju 52/3m, probably in November 1942, while *(BELOW)* is a frame from a gun camera film taken as the Beaufighters strafed the Junkers to destroy it. Flames may already be seen coming from the starboard wing. Note the tactical markings on the rudder of the Ju 52/3m, the yellow N indicating I./KGzbV 172 while the yellow spot may denote a Stab aircraft. Soon after the El Alamein breakthrough, I./KGzbV 172 was transferred to Russia to assist in the Stalingrad airlift but returned to the Mediterranean in April 1943.

A Ju 52/3m of III./KGzbV 1 over the Mediterranean. Having arrived in this theatre in April 1941, this Gruppe formed the mainstay of air transport in the area and operated throughout the North African campaign, sometimes sharing the work with only one other Gruppe. Note the Berlin Bear badge on the nose and the tactical tail code Z3F on the rudder. The latter indicates that the machine was serving with 11. Staffel when photographed, although it came from 10. Staffel.

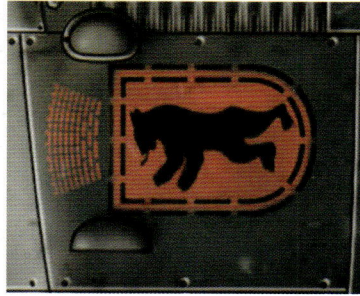

Emblem of III./KGzbV 1

Junkers Ju 52/3m 1Z+FV of III./KGzbV 1

This machine, which had an air inlet on the upper fuselage, was finished in a 70/71/65 scheme and had a white theatre band around the rear fuselage. The unit badge appeared on the nose with the shield and the individual aircraft letter F on the fuselage in the appropriate Staffel colour of red. Note that a small area of the lower right arm of the Balkenkreuz has flaked away, together with most of the 10. Staffel letter U.

LEFT: By means of special pallets attached beneath the fuselage, the Ju 52/3m could carry loads externally as well as in its cargo hold. This aircraft shown here has been loaded with a 3.7 cm anti-tank gun.

RIGHT: Supplies for the Luftwaffe in North Africa were just as important as those for the Army and sometimes ran critically short. This photograph shows Luftwaffe personnel with fuel drums and bombs which have been airlifted into North Africa by the Ju 52/3m seen in the background. Note the bombs have been packed in wooden frames which also act as sledges.

RIGHT: A decorated gantry used during engine changes marks the 15,000th sortie flown by III./KGzbV 1. Unfortunately, the names of the crew responsible for this milestone in the Gruppe's history are not known.

A loose formation of Ju 52/3ms of Transport Staffel/II. Fliegerkorps, probably in November or December 1942. At this time, the Staffel was based at Reggio, but also used Trapani, and was employed on supply flights to North Africa. When new aircraft were assigned to the Staffel, the unit badge was applied immediately although it usually took some time to replace the factory lettering. Nevertheless, the two aircraft nearest the camera have operational markings, the machine in the foreground being C3+KH which carried the tactical code S1K on the rudder.

Junkers Ju 52/3m g4e C3+KH of Transport Staffel/II. Fliegerkorps, November 1942

Although finished in a standard 70/71/65 scheme, the uppersurface colours on this aircraft were faded and the splinter pattern was indistinct. The Staffel badge appeared on the nose and the undersurfaces of the engine cowlings were yellow. The letter K of the operational markings was repeated on the leading edge of each wing, and the small white spot below the fresh air inlet on the top of the fuselage was probably a datum point. A white band encircled the rear fuselage and the tactical code S1K on the rudder was also white. Note that despite having the same operational code as the machine shown on Page 89, this is a different machine.